Environmental Applications of Chemometrics

ACS SYMPOSIUM SERIES 292

Environmental Applications of Chemometrics

Joseph J. Breen, EDITOR
Office of Toxic Substances
U.S. Environmental Protection Agency

Philip E. Robinson, EDITOR
Office of Toxic Substances
U.S. Environmental Protection Agency

Developed from a symposium sponsored by
the Division of Environmental Chemistry
at the 188th Meeting
of the American Chemical Society,
Philadelphia, Pennsylvania,
August 26–31, 1984

American Chemical Society, Washington, D.C. 1985

SEP / AE
CHEM
5287591X

Library of Congress Cataloging in Publication Data
Environmental applications of chemometrics.
  (ACS symposium series, ISSN 0097–6156; 292)

"Developed from a symposium sponsored by the
Division of Environmental Chemistry at the 188th
meeting of the American Chemical Society, August
26–31, 1984," in Philadelphia, Pa.

  Includes bibliographies and index.

  1. Chemistry—Mathematics—Congresses.
2. Chemistry—Statistical methods—Congresses.
3. Environmental chemistry—Congresses.

  I. Breen, Joseph J., 1942–   . II. Robinson, Philip
E., 1948–   . III. American Chemical Society.
Division of Environmental Chemistry. IV. American
Chemical Society. Meeting (188th: 1984: Philadelphia,
Pa.) V. Series.

QD39.3.M3E58   1985   628.5'028   85–22878
ISBN 0–8412–0945–6

ACS Symposium Series

M. Joan Comstock, *Series Editor*

Advisory Board

FOREWORD

The ACS SYMPOSIUM SERIES was founded in 1974 to provide a medium for publishing symposia quickly in book form. The format of the Series parallels that of the continuing ADVANCES IN CHEMISTRY SERIES except that, in order to save time, the papers are not typeset but are reproduced as they are submitted by the authors in camera-ready form. Papers are reviewed under the supervision of the Editors with the assistance of the Series Advisory Board and are selected to maintain the integrity of the symposia; however, verbatim reproductions of previously published papers are not accepted. Both reviews and reports of research are acceptable, because symposia may embrace both types of presentation.

CONTENTS

PREFACE

Environmental applications of chemometrics are of interest because of the concern about the effects of chemicals on humans. The symposium upon which this book is based served as an important milestone in a process we, the editors, initiated in 1982. As members of the Environmental Protection Agency's Office of Toxic Substances (OTS), we have responsibilities for the acquisition and analysis of human and environmental exposure data in support of the Toxic Substances Control Act. OTS exposure studies invariably are complex and range from evaluating human body burden data (polychlorinated biphenyls in adipose tissue, for example) to documenting airborne asbestos levels in schools.

The proper conduct of complex exposure studies requires that the quality of the data be well defined and the statistical basis be sufficient to support rule making if necessary. These requirements, from study design through chemical analysis to data reduction and interpretation, focused our attention on the application of chemometric techniques to environmental problems.

In the fall of 1982, OTS and the Agency's Office of Research and Development's (ORD) Environmental Monitoring Systems Laboratory (Research Triangle Park, NC) hosted a 2-day workshop for researchers active in chemometrics. The participants represented various agency program offices and ORD laboratories, as well as researchers from the National Fisheries and Wildlife Service, Columbia, MO; University of Illinois, Chicago; and Infometrix, Seattle, WA. It was evident that isolated attempts were in progress to apply chemometric techniques to complex environmental problems. What was lacking was a coherent chemometrics program with well-defined objectives.

The advent of analytical techniques capable of providing data on a large number of analytes in a given specimen had necessitated that better techniques be employed in the assessment of data quality and for data interpretation. In 1983 and 1984, several volumes were published on the application of pattern recognition, cluster analysis, and factor analysis to analytical chemistry. These treatises provided the theoretical basis by which to analyze these environmentally related data. The coupling of multivariate approaches to environmental problems was yet to be accomplished.

This multivariate data analysis challenge is aggressively being met by a number of researchers. The result is a vibrant and growing literature filled with software acronyms such as ARTHUR, SIMCA, CHEOPS, CLEOPATRA,

EIN*SIGHT, and others. All of these programs are specifically directed toward the multivariate analysis of analytical chemical data both in assessing data quality (quality control and quality assurance) and in interpreting data to provide insight into the complex system under investigation.

The fall of 1983 also saw the North Atlantic Treaty Organization host an Advanced Studies Institute in Cosenza, Italy, entitled "Chemometrics: Mathematics and Statistics in Chemistry." One hundred scientists—a most unusual collection of chemists, engineers, and statisticians from academia, industry, and government—representing a dozen countries assembled to discuss the role of sophisticated multivariate statistics in the daily routine of an analytical chemistry laboratory.

With this backdrop, we approached the ACS Division of Environmental Chemistry with the request to sponsor a symposium on the application of chemometrics to environmental problems.

This volume represents a majority of the presentations made at the symposium. The broad range of topics can be seen in the table of contents. Thought-provoking discussions at the symposium revealed that significant progress has been made in the application of chemometrics to environmental problems.

DISCLAIMER

This book was edited by Joseph J. Breen and Philip E. Robinson in their private capacity. No official support or endorsement by the U.S. Environmental Protection Agency is intended or should be inferred.

JOSEPH J. BREEN
PHILIP E. ROBINSON
Office of Toxic Substances
U.S. Environmental Protection Agency
Washington, DC 20460

Soft Independent Method of Class Analogy
Use in Characterizing Complex Mixtures and Environmental Residues of Polychlorinated Biphenyls

D. L. Stalling[1], T. R. Schwartz[1], W. J. Dunn III[2], and J. D. Petty[1]

[1] Columbia National Fisheries Research Laboratory, U.S. Fish and Wildlife Service, Columbia, MO 65201
[2] Health Sciences Research Center, Department of Medicinal Chemistry and Pharmacognosy, University of Illinois at Chicago, Chicago, IL 60612

Pattern recognition studies on complex data from capillary gas chromatographic analyses were conducted with a series of microcomputer programs based on principal components (SIMCA-3B). Principal components sample score plots provide a means to assess sample similarity. The behavior of analytes in samples can be evaluated from variable loading plots derived from principal components calculations. A complex data set was derived from isomer specific polychlorinated biphenyl (PCBS) analyses of samples from laboratory and field studies. The application of chemometrics to these problems includes three segments: analytical quality control; method and data base development; and modeling Aroclor composition and PCB residues in bird eggs.

Chemometrics, as defined by Kowalski (1), includes the application of multivariate statistical methods to the study of chemical problems. SIMCA (Soft Independent Method of Class Analogy) and other multivariate statistical methods have been used as tools in chemometric investigations. SIMCA, based on principal components, is a multivariate chemometric method that has been applied to a variety of chemical problems of varying complexity. The SIMCA-3B program is suitable for use with 8- and 16-bit microcomputers.

Four levels of pattern recognition have been defined by Albano (2). Levels I and II are most frequently used to determine the similarity of objects, or to characterize clusters of samples and to classify unknown objects. Level III takes advantage of the reduction of data dimensions resulting from SIMCA and seeks to establish a correlation of sample scores with independent variables

0097-6156/85/0292-0001$06.00/0
© 1985 American Chemical Society

such as chemical functions or variables, spectroscopic data or chemical toxicity. This approach is often used in quantitative structure-activity relationships (3-5). Level IV is most frequently applied to complex spectroscopic calibration problems and in situations where composition prediction or estimation is to be made from spectroscopic data.

The SIMCA approach can be applied in all of the four levels of pattern recognition. We focus on its use to describe complex mixtures graphically, and on its utility in quality control. This approach was selected for the tasks of developing a quality control program and evaluating similarities in samples of various types. Principal components analysis has proven to be well suited for evaluating data from capillary gas chromatographic (GC) analyses (6-8).

Analytical quality control (QC) efforts usually are at level I or II. Statistical evaluation of multivariate laboratory data is often complicated because the number of dependent variables is greater than the number of samples. In evaluating quality control, the analyst seeks to establish that replicate analyses made on reference material of known composition do not contain excessive systematic or random errors of measurement. In addition, when such problems are detected, it is helpful if remedial measures can be inferred from the QC data.

Our progress in the application of chemometrics to capillary GC data was advanced by the development of a laboratory chromatography data base (9). This development followed from our decision to use capillary GC in most of our laboratory analyses for environmental contaminants. A data base was considered necessary because large amounts of data were being generated from the analysis of laboratory and field studies on complex mixtures of organochlorine contaminants. A data base is an important, but not essential, factor in using pattern recognition for quality control.

The most advanced application of pattern recognition (Level IV) offers the possibility of predicting independent variables by using latent variables derived from examining training sets of dependent and independent variables (10). The application of partial least squares in the prediction of the composition of mixtures of Aroclors was previously explored (6) by using the program, PLS-2 provided by the SIMCA-3B programs (11-12).

The first results from the use of PLS were reported by Dunn et al (6) who estimated the composition of PCB contaminated waste oil in terms of Aroclor mixtures. Stalling et al (13), who reported on the characterization of PCB mixtures and the use of three-dimensional plots derived from principal components, demonstrated that the fractional composition of TCDD and other PCDD residues were related to their geographical origins. These two reports (6,13) described the application of an advanced chemometric tool in residue studies and illustrated the use of pattern recognition to extract quantitative information about sample similarity.

In our present investigations, we encountered a pressing need for an objective, statistically based way of evaluating concentrations of as many as 105 individual PCB isomers in each sample analyzed by capillary GC. We summarize here some of the experience obtained in our laboratories from the use of SIMCA to characterize Aroclor mixtures and environmental PCB residues in a series of bird eggs.

METHODS

Sampling. Eggs of Forster's tern (Sterna forsteri) were collected in 1983 from nests in two colonies in Wisconsin--one on Lake Poygan and the other on Oconto Marsh, Green Bay--as part of a study on impaired reproduction. Lake Poygan is a relatively clean lake whereas Green Bay is heavily contaminated from the Fox River with many industrial chemicals--particularly PCBs and chlorophenols, which are known sources of PCDFs and PCDDs. Reproductive success has declined and the incidence of deformed young has increased in the Green Bay colony (14).

Analysis of PCBs. PCB residues in extracts of egg samples were enriched by using a combination of gel permeation chromatography on BioBeads S-X3 and 1:1 (v/v) cyclohexane:methylene chloride. Adsorption column chromatography on silicic acid was used to separate PCBs from other co-extractives and contaminants (15).

The PCB congeners were separated by using a glass capillary chromatographic column (30 M x .25 mm i.d.) coated with C_{87}-hydrocarbon stationary phase (Quadrex Corp., New Haven, CT 06525); a 60-cm uncoated fused silica retention gap connected the injector to the analytical column and a 15 cm uncoated fused silica column connected analytical column to the detector. The data sampling and gas chromatography program was controlled by a Varian Autosampler Model 8000, which also delivered a calibrated amount of sample to the GC injection port. Chromatographic conditions were similar for all of the analyses: initial temperature, 80 °C, programmed at 3 °C/min to a final temperature of 265 °C; detector temperature, 320 °C; and injector temperature (direct inject) 220 °C.

An IBM CS9000 microcomputer was interfaced with the GC which acquired data generated by the electron capture detector. In processing the data, we used the CS9000 and a software package designed for laboratory data collection (Capillary Applications Program [CAP], IBM Instrument division, Danbury, CT 06810). We organized the processed peak data, using a basic program, into a series of files on hard disk media and transferred these files off-line to a Digital Equipment Corp. (DEC) PDP-11/34 minicomputer. We

then organized the data into tree-structured disk files,
using our specialized laboratory data base management
computer programs written in DSM-11 (Digital Standard
MUMPS) for the PDP-11 family of computers.

We separated 105 constituents and achieved
calibration by using a 1:1:1:1 (w/w/w/w) mixture of
Aroclors 1242, 1248, 1254 and 1260. The last two digits of
the Aroclor number designates the percentage chlorine in
the Aroclor. A chromatogram of this mixed Aroclor
standard is shown in Figure 1. The method of peak
identification was a retention index system utilizing n-
alkyl trichloroacetates (16). Molar response factors were
determined from a flame ionization detector by using the
computer-based calculation methods described by Schwartz
et al. (16).

After we determined the concentrations of individual
isomers, we retrieved the data from the MUMPs based
laboratory data base, and transferred them to an IBM-XT
(IBM Corporation, Boco Raton, FL 33432) by way of a RS-232
link, using the program Cyber (Department of Linguistics,
University of Illinois at Champaign-Urbana, Urbana, IL
61820). In performing principal components analyses, we
used SIMCA-3B for MS-DOS based microcomputers (Principal
Data Components, 2505 Shepard Blvd., Columbia, MO 65201).

A series of Aroclors and known Aroclor mixtures were
analyzed by these techniques to provide a training data
set for SIMCA-3B. These standards included replicate
analyses, a 1:1 (w/w) mixture of each Aroclor in
combination with one other Aroclor, and a 1:1:1:1 mixture
of each Aroclor (Table I).

Principal Components Analysis

We examined the data by calculating principal components
sample scores (Thetas) and variable loading terms (Betas),
using the program CPRIN from the SIMCA-3B programs. After
calculating two or three principal components for a class
model, one can prepare a plot of sample similarity, by
using the sample scores (Theta-1 vs Theta-2), as well as
variable loadings (Beta-1 vs Beta-2). Sample similarity
was determined by calculating sample scores (θ-values,
Equation [1]).

$$X_{ik} = \overline{X}_i + \sum_{a=1}^{A} \theta_{ka} \cdot B_{ai} + E_{ik} \qquad [1]$$

The likeness of samples within the class can be
assessed by the proximity of samples to each other in
plots derived from principal components models. The
statistical technique of cross-validation (17) was used to

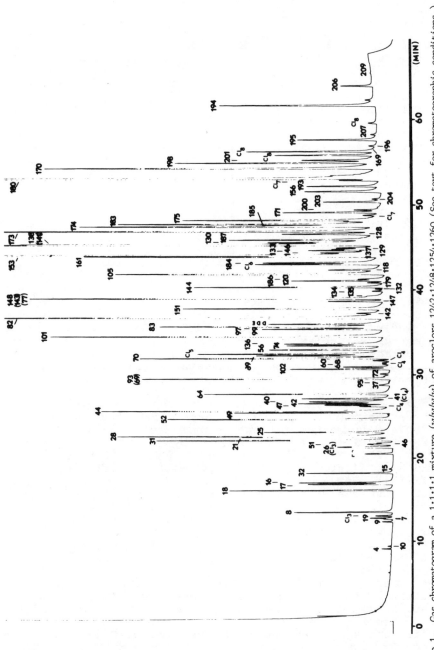

Figure 1.   Gas chromatogram of a 1:1:1:1 mixture (w/w/w/w) of aroclors 1242:1248:1254:1260 (See text for chromatographic conditions.)

determine the number of components that were statistically significant.

Table I. Aroclors Samples Composing the Training Data Set.

	Aroclor Composition				
Sample #	1242	1248	1254	1260	Replicate #
1	0	0	1	0	1254-1
2	0	1	0	0	1248-1
3	0	0	0	1	1260-1
4	1	0	0	0	1242-1
5	1	0	0	0	1242-2
6	0	0	0	1	1260-2
7	0	0	1	0	1254-2
8	1	1	1	1	1:1:1:1-1
9	1	1	1	1	1:1:1:1-2
10	1	0	1	0	-
11	1	1	0	0	-
12	0	1	1	0	-
13	0	0	1	1	-
14	0	1	0	1	-
15	1	0	0	1	-

Principal Components Plots

By using SIMCA-3B program, FPLOT.EXE, one can plot numerous variables derived from the principal components calculations. Because a printer in the character mode is used with this program to plot variables, the plots are restricted to two-dimensional presentations.
    The program 3DPC.BAS (Principal Data Components) provides a means to plot sample scores in 3-D and color if three principal components are calculated. The 3-D display derived from the sample score values may be transferred to a disk file by using the program, FRIEZE.COM, supplied as part of PC-PAINT BRUSH or 4-Point Graphics (International Microcomputer Software, Inc., [IMSII]), San Rafel, CA 94991). The image is stored on disk and can be edited, enhanced, or labeled with a commercial software package such as PC Paint Brush (IMSII). The screen image can also be printed on a color or black/white printer.

RESULTS and DISCUSSION

Analyses of PCBs can create large data sets that are difficult to interpret, since there are 209 PCB isomers. Isomer compositions may vary widely due to differential partitioning or metabolism of compounds. In addition, wide differences in residue profiles may exist in the biota locally because of variations in effluents, combustion, or other source of residues. Chemometric methods can greatly improve the analyst's ability to describe and model residues in these diverse samples.

The utility of principal components modeling of multivariate data like those encountered in these complex mixtures, originates from graphical presentations of sample similarity, as well as from statistical results calculated by the SIMCA-3B programs (3). Sample data are treated as points in higher dimensional space, and projections of these data are made in two- or three-dimensional space in a way that preserves most of the existing relations among samples and variables (3). This feature is especially helpful in visualizing data of more than three dimensions.

The calculations involved in principal components are summarized in Equation [1]. The objective was to derive a model of a data set having $\underline{k}$ samples and $\underline{i}$ variables in which the concentration or value of any measured value, X_{ik}, could be calculated. The principal component term is the product of Θ_{ka} and B_{ai}, where Θ_{ka} (Theta) is designated the ath component "score" for sample k, and B_{ai} (Beta) is designated as the "loading" for variable $\underline{i}$ in principal component $\underline{a}$. The term $\overline{X}_i$ is the mean of variable X_i in all samples. The residual term (or unexplained part of the measurement not modeled) is designated E_{ik}, and "A" describes the number of principal components extracted from the data. A more detailed discussion of this approach was given by Dunn et al. (6,18).

The concentration data obtained from each sample analysis were expressed as fractional parts and normalized to sum to 100. The normalized data were statistically analyzed, and three principal components (A=3, Equation [1]) were calculated. The PCB constituents (varibles) are numbered sequentially and correspond to peak #1, peak #2, ... to peak #105. The structure and retention index of each constituent in the mixture were reported by Schwartz et al. (9). The tabular listing of the data is available from the present authors.

The Aroclor samples listed in Table I were modeled by principal components to illustrate how the result from principal components calculations can be used in describing PCB data. The sample scores (Figure 2, A.- Theta-1 vs. Theta-2; B. Theta-1 vs. Theta-2; and C.- Theta-2 vs. Theta-3) are plotted for the samples.

Results obtained from the plots of the variable loadings (Figure 2, A'- C') for the three components provide insight into the importance of the GC Peaks in separating the various Aroclors and their mixtures (Figure 2, A - C). The loading plots show a separation of variables that are tightly clustered, the groups of variables radiating outward from the center. They are clustered in groups that reflect the variables that are characteristic of the individual Aroclors.

The sample scores (Theta-1, Theta-2, and Theta-3) in each component were used to represent the samples in a 3-D

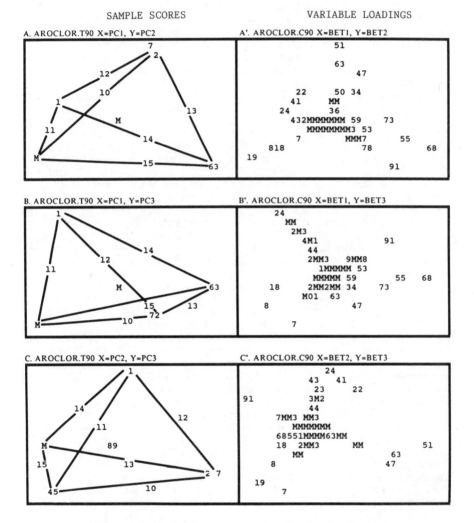

*Figure 2. Principal Components Plots for Aroclor Samples
(ref. Table 1 for Sample i.d.)*

graph (Figure 3). This plot shows that while much of the
sample information may be discerned from a two component
model, it is impossible to tell if an Aroclor mixture is
composed of more than a mixture of three Aroclors. The 3-
D presentation illustrates more clearly than three 2-D
plots, how complex data may be viewed and relations among
the samples more clearly comprehended than when data are
presented in tabular form.

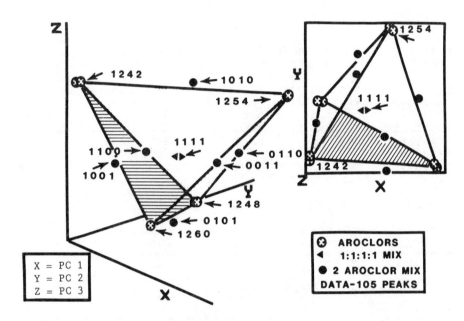

Figure 3. *3-D Plot of Principal Components Scores (Theta-1,-2,-3) Representing Normalized Isomer Composition Data for Aroclors 1242, 1248, 1254, 1260, and their mixtures. The points for each Aroclor represent individual sample analyses. The plot in the upper right quadrant is the view parallel to the Z-axis.*

The principal components model of the Aroclor samples
(Table I)   preserves greater than 95% of the sample
variance of the entire data set. From the 3-D sample
score plot (Figure 3) one can make these observations: PCB
mixtures of two Aroclors form a straight line; three
Aroclor mixtures form a plane; and that possible mixtures
of the four Aroclors are bounded by the intersection of
the four planes. Samples not bounded by or inside the
volume formed by the intersection of the four planes may

be derived from, but are not identical to, mixtures of Aroclors.

In SIMCA-3B, modeling power is defined to be a measure of the importance of each variable in a principal component term of the class model (18). The modeling power has a maximum value of one (1.0) if the variable is well described by the principal components model. Variables with modeling power of less than 0.2 can be eliminated from the data without a major loss of information (18).

For the PCB mixtures we analyzed (Table I), the modeling power was determined on the basis of a three component model (A=3). These data revealed that most of the 105 GC-peaks play an important role in the class model for the four Aroclors and their mixtures. The modeling power of each variable is plotted (Figure 4) for each component term along with a plot of the concentration profile of sample 9 (Table I); this sample contains 1242:1248:1254:1260 in a 1:1:1:1 ratio and the plot represents its fractional composition.

Reproducability aspects of the analysis is reflected in the nearly identical proximity of each of the replicate analyses (Table I). Use of this statistical technique to examine sample residue profiles from different locations has lead to an improved understanding of complex mixtures of contaminants and related problems.

Residues in Forster's Tern Eggs

A decline in reproductive success and increased incidence of deformed young were observed in colonies of Forster's terns, common terns, cormorants, and herons in and near Green Bay (14, 20). The samples from two Wisconsin locations (Lake Poygan and Green Bay) were analyzed for individual PCBs using the method described by Schwartz et al. (9). Residue levels for the total PCB content (Table II) represent the sum of the individual PCBs present in the sample.

For the SIMCA analyses, the individual PCB isomer concentrations were normalized to sum 100. We examined the data by using the SIMCA-3B program to calculate principal components and to plot sample scores in a manner identical to that discussed for the Aroclor mixtures. The plot of sample data illustrates that the geographic locations have different residue profiles (Figure 5).

The PCBs present in eggs collected from Green Bay birds were similar to Aroclor 1254, and the total PCB concentrations were about 6 times greater than those in Lake Poygan. The composition of PCBs in eggs collected from Lake Poygan birds were less consistent and tended to lie farther from a line between Aroclors 1254 and 1260 (Figure 5). We found that the geographical origin of the samples (Lake Poygan or Green Bay) could be ascertained with a probability of 0.85 by using a class

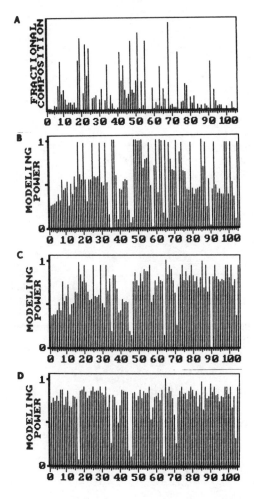

Figure 4. Plots of the Fractional Composition of an Aroclor Mixture (A, Sample 9, Table 1) and the Modeling Power for a Three Component Model of the Samples in Table 1: PC-1 (B); PC-2 (C); and PC-3 (D).

model of each group based on normalized residue data. This was determined using the program the SIMCA-3B program "CLASSI" to classify the samples.

Table II.   Residues of PCB in Tern Eggs Collected from Lake Poygan
            and Green Bay1

Collection Site	Mean PCB Residue (ug/g wet weight)
Lake Poygan	20.4 (7.8) [2]
Green Bay	3.7 (1.5)

[1]each composite sample contained 6 eggs
[2]standard deviations in parentheses

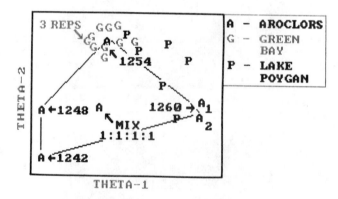

Figure 5. Principal Components Plot Derived from Analysis of Aroclor
Standards, Their Equal Mixture, and Eggs of Forster's Tern.

Two Aroclor 1260 standards (A_1 and A_2) were included in these analyses.   One standard was from the Columbia National Fisheries Research Laboratory, and the other from the Patuxent Wildlife Research Center (U.S. Fish and Wildlife Service, Laurel, MD.)   A difference in the concentration of one constituent of about 30% was responsible for the small difference observed between the two Aroclor 1260 standards (Figure 5.)   Use of a quantitative chemometric method to describe compositional residue differences measured in environmental samples may prove helpful in correlating residue profiles and concentrations with observed biological effects, such as decreased survival of young birds.

SUMMARY

These applications demonstrate that pattern recognition techniques based on principal components may be effectively used to characterizate complex environmental residues. In comparisons of PCBs in bird eggs collected from different regions, we demonstrated through the use of SIMCA that the profiles in samples from a relatively clean area differed in concentration and composition from profiles in samples from a more highly contaminated region. Quality control can be evaluated by the proximity of replicate analysis of samples in principal components plots.
More extensive use of isomer specific analysis, when combined with chemometric techniques, should improve insight into how residues in the environment relate to their sources. This approach could lead to a quantitative description of changes in the composition of these chemicals as they pass through the food chain and are distributed in the environment.

Acknowledgment

We thank Michael Koehler (University of Illinois at Chicago) for developing the combined two- and three-dimensional plotting program to display the sample component scores.

Disclaimer. References to trade names or manufacturers of commercial products do not imply Government endorsement of commercial products.

Literature Cited

1.   Kowalski, B.; Chemistry and Industry, 1978, 18, 882-884.

2.   Albano, C.; Dunn, W.J., III; Edlund, U.; Johansson, E.; Nroden, B; Sjostrom, M.; Wold, S. Anal. Chim. Acta. Comp. Tech. Optim. 1978, 103, 429-443.

3.   Wold, S.; Sjostrom, M. In "Chemometrics, Theory and Application," ACS Symp. Ser. 1977, No. 52, 243-282.

4. Dunn, W. J., III; Wold, S. J. Med. Chem. 1979, 21,
    1001.

5. Dunn, W. J., III; Wold, S. J. Chem. Inf. Comput. Sci.
    1981, 21, 8-13.

6. Dunn, W. J., III; Stalling, D.L.; Schwartz, T.R.;
    Hogan, J.W.; Petty, J.D.; Anal. Chem. 1984, 56,
    1308-1313.

7. Stalling, D.L.; Smith, L.M.; Petty, J.D.; Dunn, W.J.,
    III, "Dioxins and Furans in the Environment -- A
    Problem for Chemometrics," in "Dioxins in the
    Environment," Limno-Tech, Inc., Ann Arbor, MI, in
    press.

8. Stalling, D.L.; Dunn, W. J., III; Schwartz, T. R.;
    Hogan, J. W.; Petty, J. D.; Johansson, E.; Wold, S. In
    "Application of SMICA, A Principal Components Method,
    in Isomer Specific Analysis of PCB's," ACS Symp. Ser.
    1985, in press.

9. Schwartz, T. R.; Campbell, R.D.; Stalling, D.L.;
    Little, R.L.; Petty, J.D.; Hogan, J.W.; Kaiser, E.M.;
    Anal. Chem. 1984, 56, 1303-1308.

10. Wold, S. and Dunn, W. J., III.; J. Chem. Inf. Comput.
    Sci., 1983, 23, 6-13.

11. Wold, S.; Pattern Recognition, 1976. 8, 127-134.

12. Wold, S., Albano, C., Dunn, W.J., III, Edlund, U.,
    Esbensen, K. Geladi, P., Hellberg, S., Johansson, E.,
    Lindberg, W., and Sjostrom, M., "Multivariate Data
    Analysis in Chemistry," in Chemometrics. Mathematics
    and Statistics in Chemistry, Kowalski, B.R., Ed., D.
    Reidel Publishing Company, 1984, 17-95.

13. Stalling, D.L., Norstrom, R.J., Smith, L.M., Simon,
    M., Chemosphere, 1985. in press.

14. Memorandum from T.J. Kubiak, Fish and Wildlife
    Service, Green Bay Field Office, University of
    Wisconsin-Green Bay, Green Bay, WI, to Members Fox
    River/Green Bay Toxics Task Force, January 27, 1983.

15. Stalling, D.L.; Tindle, R.C.; J. Assoc. Off. Anal.
    Chem., 1972, 55, 32-38.

16. Schwartz, T. R.; Petty, J. D.; Kaiser, E. M. Anal.
    Chem. 1983, 55, 1839.

17. Wold, S.,Technometrics 1978, 20, 397-406.

18. Wold, S., Technometrics 1978, 20, 397-406.

19. Dunn, III, W.J., Wold, S., and Stalling, D.L., "How SIMCA Pattern Recognition Works," Proceedings of Symposium on Chemometrics, Division of Environmental Chemistry, 188th National ACS Meeting, Philadelphia, PA, August 26-31, 1985. in press.

20. Harris, H.J., and J. A. Trick, 1979, Annual Performance Report, Wisconsin Department of Natural Resources, Office of Endangered and Non-Game Species, Madison, Wisconsin.

RECEIVED July 17, 1985

Evaluating Data Quality in Large Data Bases Using Pattern-Recognition Techniques

Robert R. Meglen and Robert J. Sistko

Center for Environmental Sciences, University of Colorado at Denver, Denver, CO 80202

Increased sophistication of chemical instrumentation and computerized data acquisition have quantitatively and qualitatively changed analytical chemistry. Chemists measure more variables and perform more experiments in less time than feasible just a few years ago. Without a concomitant enhancement of interpretive skills the new-found data affluence may be a curse and not a blessing. More data tend to cloud the issue rather than clarify it. The result of the paradox is that many useful observations remain uninterpreted. Pattern recognition techniques have been used to enhance the interpretation of large data bases. This paper describes how these techniques were used to examine a large water quality monitoring data base. The paper describes how pattern recognition techniques were used to examine the data quality, identify outliers, and describe underground water chemistries.

Intensive instrumental and analytical methods research performed during the 1970's has clearly contributed to the confidence with which current research results are reported. Examination of recent literature shows that research protocols have departed from simplistic single element studies and have incorporated more realistic experimental designs that include multi-elemental determinations. This change reflects a growing awareness that chemical interactions between chemical species are important in complex chemical systems. Increased reliance on multi-elemental analysis reflects the ease with which such analyses can be performed. Recent advances in electronics, chemical instrumentation, and computerized data acquisition have quantitatively and qualitatively changed analytical chemistry. Chemists measure more variables and perform more experiments in less time than feasible just a few years ago. In spite of our recently acquired data affluence, many complex problems remain unsolved. The enhanced insight that additional data were to provide has failed to materialize. In some cases, more data cloud the issue rather than clarify it. Acquiring massive quantities of data is ineffective until interpretations are made and incorporated into a mechanistic description of the system.

0097-6156/85/0292-0016$06.00/0

Data are not information. Powerful interpretive aids that match
the sophistication of the instrumental tools are required to
facilitate the task of converting data into information and finally
into knowledge of the system. In recent years an increasing number of
"data burdened" researchers (1-3,6,9-14,16,18) have begun to employ
some of the statistical techniques that found broad application in
the social sciences during the 1940's. These techniques, loosely
termed pattern recognition, are based upon factor analysis, principal
component analysis, classification analysis and cluster analysis.
These techniques greatly enhance the assimilation of massive data
bases and provide a valuable mechanism for summarizing multivariate
data. As an illustration of how these techniques assist in the examination of large data bases we shall employ an example from the
field of environmental chemistry.

Environmental studies are often characterized by large numbers
of variables measured on many samples. When poor understanding of the
system exists one tends to rely upon the "measure everything and hope
that the obvious will appear" approach. The problem is that in complex chemical systems significant patterns in the data are not always
obvious when one examines the data one variable at a time. Interactions among the measured chemical variables tend to dominate the data
and this useful information is not extracted by univariate approaches.

The need for multivariate techniques is apparent when one con-
siders that each measured parameter contributes one dimension to the
representation. Thus examining two parameter interactions requires a
two dimensional plot. Such graphical representations are effective in
identifying significant relationships among the variables. A three
variable system requires a three dimensional plot to simultaneously
represent all potential bivariate interactions. However, as the
number of variables increases the dimensionality of the required
representation exceeds man's ablilty to perceive significant patterns
in the data. Indeed, humans do not conceptualize comfortably beyond
three dimensions. Without assistance one would be restricted to
considering only problems that are characterized by three factors.
(If one restricts the interpretive task to two variable interactions
one may generate a series of two dimensional graphs, one for each
unique bivariate pair. Again, the mere task of examining all of the
plots becomes formidable. A data base consisting of 35 measured
variables would require examining 595 plots!) One commonly computes a
correlation matrix consisting of all unique bivariate correlation
coefficients to summarize the variable interactions. While this type
of summary is helpful, it provides little insight regarding the
natural associations among groups of variables. The more powerful
factor analytic treatment extracts the significant underlying
relationships that characterize the data. Factor analysis provides
the tools by which data are converted to information. It is in these
natural associations that one hopes to find the clues to uncover
otherwise obscure mechanisms.

A second capability that one needs in examining large data bases
is a convenient way to represent relationships among samples or
objects upon which the measurements have been made. This procedure is
analogous to the search for variables that are associated with one
another. Group behavior among the objects indicates that significant

distinctions are possible, and the distinctions lead to useful
generalizations that simplify complex systems. In addition to providing a useful summary of the total data base, this representation
provides a valuable aid to identifying unusual sample behavior. Once
unusual behavior has been identified, one can begin to examine the
possible causes. The causes for anomalous behavior are often simple
measurement error. Thus, identifying outliers helps focus attention
on the distinctions that make a difference.

The Rationale for Using Pattern Recognition

We will illustrate the application of pattern recognition techniques
on a water quality data base. Baseline water quality data was acquired by Cathedral Bluffs Shale Oil Company and supplied to our
laboratory on magnetic tape. The water quality monitoring program on
oil shale lease tract C-b (western Colorado) was designed to comply
with State permitting requirements. The data had not been examined
previously because the data base suffered from many of the same
limitations described earlier. The monitoring system consisted of six
different types of waters as illustrated in Figure 1. Monitoring
wells were drilled to permit sampling deep bedrock aquifers above and
below an aquitard (the Mahogany Zone) which was to be mined for its
rich oil shale. It had been known that the upper and lower aquifers
contained waters with different qualities. One of the goals of the
monitoring program was to determine whether the upper and lower
aquifers communicate with one another, and whether future mining
within the aquitard separating them might introduce communication
that could degrade aquifer water quality. Monitoring wells drilled
into the upper aquifers were designated as "WX" wells and wells
sampling the lower aquifers were designated as "WY" wells. Shallower
monitoring wells, designated as "WA's", were drilled to sample waters
contained in the unconfined aquifers of the alluvium. Other surface
waters, springs and seeps, in contact with the alluvium were also
sampled. These were designated as "WS"-type waters. Another goal of
the monitoring program was to determine whether any WA or WS type
waters originated in either of the deeper confined aquifers (WX and
WY). Water seeping into the mined zone was pumped to suface holding
ponds. Shallow wells ("WW"-type) drilled into the alluvium around the
holding ponds were monitored to detect any leaks in the holding
ponds. Several WX and WY wells were recompleted during the monitoring
program in order to sample specific regions of the upper and lower
aqufers. These are designated as R wells. Thus water samples from 6
categories and 89 sampling sites were analyzed for 35 chemical
parameters over a five year period. Prodigious numbers of conventional two dimensional plots would be required to examine parameter
versus time and all other bivariate relationships.
    Simple examination of plots of each variable versus time would
require 35 plots. If distinction among the different sampling locations were to be examined 3115 plots would be required. Since it is
usually of interest to determine whether there are any significant
bivariate relationships (correlations) among the measured variables,
one would need to examine a total of 595 scatter plots. If one were
to seek discrimination among each of the categories 3570 plots would
be required. In a monitoring program it is important to examine the
individual behavior of each individual sampling site; this requires

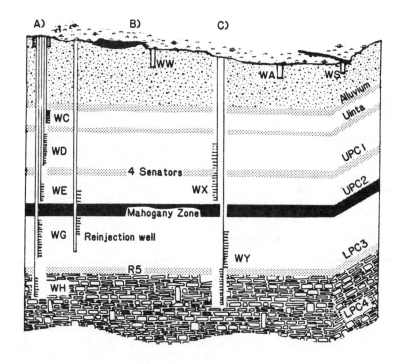

Figure 1. Schematic diagram showing the monitoring system. (A) recompleted well, (B) surface holding pond with WW monitor, (C) WX/WY deep bedrock wells.

52,955 plots! Thus, complete examination of the data base would
require the preparation and examination of over 60,000 unique plots.
It is likely that only a few of these plots would yield insight about
the relationship of the measured variables with one another and the
underlying geochemistry of the waters, but we need an unbiased guide
to decide which plots to examine. In addition, even if we were to
examine all plots we could still miss relationships that involve more
than two variables at a time.

In addition to the quantitative difficulties attending conventional data base examination there are several qualitative limitations imposed by the nature of the measurements themselves. Many
standard statistical techniques require some knowledge or assumptions
about the shape of the distribution of measured values. Many environmental variables do not have well behaved distributions; some are
highly skewed and some are multimodal. Robust analysis of these data
requires techniques that do not rely upon a priori knowledge or
assumptions about the underlying variable distributions. Qualitative
limitations and the magnitude of the interpretive task mandates a
computer assisted examination of all possible relationships. We have
exploited the power of computer assisted pattern recognition techniques in examining this data base.

The mathematical techniques employed in pattern recognition
permit rapid and efficient identification of relationships and key
aspects that otherwise might remain hidden in the large mass of
numbers. Since the data base was not well characterized we set the
following objectives for the interpretive study:

Evaluate the quality of the data.

Describe the water chemistries represented in the sample
categories.

Determine the key chemical parameters that govern non-stochastic
behavior.

Determine the "chemical fingerprints" that identify the various
waters.

Develop a classification model that permits aquifer indentifica-
tion.

Identify the sampled aquifers and determine whether interaquifer
communication occurs.

Pattern Recognition Principles

There are several advantages that obtain from the application of
pattern recognition for data interpretation. The methods can rapidly
identify key variables that are important to time related changes in
the monitoring wells or among the well categories. This greatly
reduces the number of two dimensional plots that must be examined
since the technique extracts the relatively few plots that are likely
to be most effective in displaying differences that make a difference. Another advantage of these techniques is that they are
multivariate, they incorporate many variables simultaneously. Since

complex chemistry is involved one must rely upon interpretive techniques that are sensitive to variable interactions. Since the factor
analysis technique begins by computing the variance-covariance matrix
it incorporates information about subtle changes that slightly affect
several variables simultaneously. This type of change is easily
missed when examining only one or two variables at a time. In addition, the technique gives equal weight to variables with small absolute values relative to other large variables. This is accomplished
by autoscaling or z-scoring data. When one examines unscaled data
there is a natural tendency to focus on variables with large magnitudes; thus, significant information about chemical interactions
among trace elements is easily missed. Autoscaling the data ensures
that the search for relevant variance among small numbers is not
obscured by large invariant features.

The pattern recognition approach consists of two phases; exploratory data analysis and applied pattern recognition (classification model development). The purpose of the exploratory phase is
to uncover the basic relationships that exist among the variables and
between the samples. The applied pattern recognition phase tests the
strength of these basic relationships and other presumed relationships by developing classification-prediction models and determining
their accuracy. A brief description of the exploratory phase is given
here. Details of the procedures may be obtained from published work
(4,5,7,15).

Before any work can proceed it is necessary to prepare the
existing data base for computerized examination. A data cleanup step
consists of a search for several potential problems that could
restrict the full utilization of the existing data. Columns of
measurements that are incomplete can be "filled" if only a small
percentage of samples have missing measurements. It is possible to
"fill" the missing data items in an unbiased way so that all of the
measurements can be used. Several filling options are available to
accomplish this goal. However, complete treatment of these techniques
is not possible here. Missing data in the example presented here were
filled by substituting the variable's mean value for missing data
items. If most of the measurements are missing one must delete the
variable from further consideration. Chemical data are often entered
as "below detection limit". Designation as below detection limit is a
quantitative determination; i.e. it contains useful information
relative to all samples that exceed the detection limit. Protocols
for pre-treating detection limit data permit exploitation of the
information contained in these numbers. Once the data base has been
prepared for use it is placed into a storage format for the computer
algorithms (in this case we used a pattern recognition package called
ARTHUR (8) and multivariate statistical routines found in SPSS.(17))

Exploratory data analysis is designed to uncover three main
aspects of the data:

● anomalous samples or measurements

● significant relationships among the measured variables

● significant relationships or groupings among the samples

Exploratory data analysis is an iterative process in which a wide

variety of tools are employed. There is no set sequence in which
these tools are applied. Each data base may be approached in a different way, but after all of the iterations and alternate paths have
been explored the key findings should converge to a single coherent
summary of the data base. The first approach is consistent with the
most basic assumption of the exploratory analysis, that all of the
data are "good" and that nothing is known about the structure of the
data base. This approach is particularly useful when other interpretation attempts by techniques other than pattern recognition have
been exhausted. This approach is powerful since it does not impose a
bias about the data base that precludes exploring so-called unfruit-
ful paths. By initially including all of the data, regardless of any
predisposition toward its value, we rely upon the pattern recognition
algorithms to identify unusual behavior. There is a fundamental
philosophical reason for preferring this approach. Instead of searching the data for an answer, we ask the more fundamental question,
"What do the data tell us?" By examining anomalous behavior our
attention is focused on distinctions that can be made and relationships that can be identified. Thus, in each iteration we identify a
difference and then attempt to explain it. After successive layers of
explainable results (information) have been peeled away only "noise"
remains. As each anomaly is identified and confirmed by independently
established knowledge of the data base, one gains insight about the
system and confidence that useful information is being uncovered. The
older literature on artificial intelligence calls this approach
unsupervised learning. The three primary tools used in this approach
are factor analysis, principal component analysis, and cluster
analysis.

Factor Analysis

Factor analysis typically consists of two steps; a strictly mathematical step called principal component analysis, followed by a
refinement step that employs mathematical tools to enhance the inter-
pretability of the extracted factors. The aim of factor analysis is
to identify the few important dimensions (i.e., factors or "types" of
variables) that are sufficient to explain the meaningful information
in the data set.

Since each measured parameter adds a dimension to the data
representation, measurement of 35 variables requires the ability to
depict relationships in a 35-dimensional space. This is well beyond
the two or three dimensions where humans conceptualize comfortably.
It is also beyond the graphical representation capabilities commonly
used. Factor analysis is one of the pattern recognition techniques
that uses all of the measured variables (features) to examine the
interrelationships in the data. It accomplishes dimension reduction
by minimizing minor variations so that major variations may be summarized. Thus, the maximum information from the original variables is
included in a few derived variables or factors. Once the dimensionality of the problem has been reduced it is possible to depict
the data in a few selected two or three dimensional plots. We shall
see how these plots highlight the significant features of the underlying data structure.

In addition to the graphical representations we also obtain a
set of simple linear combinations of variables that enable us to

quantify similar or parallel behaviors among the measured variables. These variable groupings permit us to generalize the behaviors into factors. Qualitatively different areas where little generalization can be made between two areas are referred to as separate factors. An example of a factor might include a group of chemical elements which, upon inspection, suggests that the variables included in the factor characterize a particular mineral. Recall that these factors arise out the natural association of these elements with one another, information derived from the chemical analyses recorded in the data, not from any structure imposed by the data analyst. The natural associations among the variables is quantified by computing the correlation coefficients among all variable pairs. The technique known as principal component analysis is accomplished by the mathematical tool of eigenanalysis. Eigenanalysis extracts the best, mutually independent axes (dimensions) that describe the data set. These axes are the so-called factors or principal components. The utility of constructing a new set of axes to describe the data is that most of the total variance (information) in the data set may be concentrated into a few derived variables. This means that instead of having to depict the data on dozens of bivariate plots we can recompute the original sample measurements in the new data space and depict most of the information on just a few two dimensional graphs called factor score plots. This process may be viewed as projecting into two dimensions the original data from its multidimensional representation. As with any projection, information is lost; but this technique maximizes the retention of information and quantifies the amount of information contained within each projection. In most chemical systems it is possible to depict 80-90% of the total information in less than half a dozen plots.

The second step in factor analysis is interpretation of the principal components or factors. This is accomplished by examining the contribution that each of the original measured variables makes to the linear combination describing the factor axis. These contributions are called the factor loadings. When several variables have large loadings on a factor they may be identified as being associated. From this association one may infer chemical or physical interactions that may then be interpreted in a mechanistic sense.

Results and Use of Factor Score Plots

Once the principal component analysis on the unexpurgated data has been completed one can construct factor score plots that depict the location of all the samples. These plots facilitate identification of anomalous behavior. Figures 2 through 4 illustrate multiparameter anomalous behavior found in the groundwater quality data. The circled points in Figure 2 display anomalies that are not easily detected by conventional univariate examination. Figure 2 shows two samples at the axes extrema. When these points are identified by sample number the monitoring records may be examined for potential causes. In this case the records indicate that a data transcription error probably occurred. Figure 3 shows the projection on to the plane defined by the first and third principal components. This plot shows that a few samples spread away from the majority of samples along the vertical and horizontal axes. Examination of the data base indicates that points in the shaded region correspond to samples that were collected

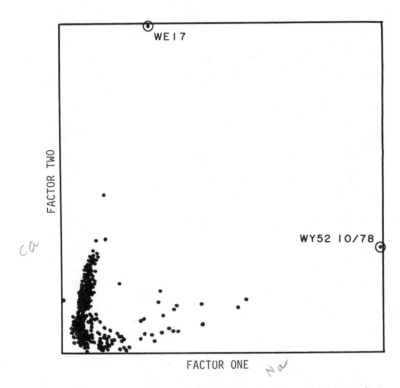

Figure 2. Factor two (hardness) vs. Factor one (salinity) factor score plot for 679 samples. Data entry errors identified.

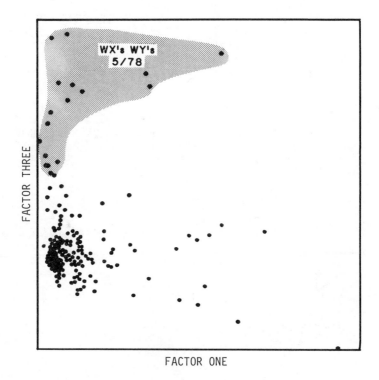

Figure 3. Factor three (metals) vs. Factor one (salinity) factor score plot for 675 samples. Sampling errors identified.

on a single day, one from each WX and WY monitoring well. This behavior suggests a sampling bias on that particular day. While the
sampling records do not permit us to ascertain the cause, it is
likely that samples collected on this day were obtained without the
usual well swabbing that nominally precedes sampling. Figure 4 illustrates, in a single graph, the existence of analytical bias between two laboratories that supplied results. The fact that the
original contract analytical laboratory was replaced during the
monitoring program was not supplied to us prior to our hypothesizing
it from the plot features. The x-axis separates the arsenic, barium,
chemical oxygen demand, lithium, pH, and strontium results obtained
by the two laboratories. The vertical axis further identifies the
difficulty that the first contract laboratory experienced in gener-
ating accurate trace metal (Mn, Fe, B, Cu, Ni, and Na) determinations
on the first samples they received. Subsequent analyses performed by
this laboratory, shown in the shaded region, at the lower right side
of the plot, agreed more closely with the trace elemental analyses
performed by the second contract laboratory during 1978 through 1981.
However, the first laboratory still produced biased analyses for As,
Ba, COD, Li, pH and Sr.

Figure 5 illustrates that not all anomalous behavior is associated with systematic bias in sampling or analytical performance.
Wells designated WW-12 and WW-13 are shallow wells drilled in the
vicinity of surface holding ponds. WW-12's behavior (movement along
Factor Four with time) indicates a leak from the holding pond. This
finding was later confirmed by Cathedral Bluffs' personnel. The pond
was relined with bentonite clay and well WW-12 was redesignated WW-
22. Samples were taken monthly for seven months. The sudden jump
along Factor Four from WW-12 to WW-22 could not be explained; the
leak had supposedly been repaired. Cathedral Bluffs offered an explanation; after the relining procedure a sampling device became
lodged in the well and a small explosive device was detonated in the
well to free the apparatus. The charge contained an ammonium salt and
this poses the best explanation of the anomalous behavior along the
axis identified by ammonium and Kjeldahl nitrogen. This example
illustrates the technique's power to identify anomalies that would
not be identified by conventional data treatments. It further illustrates that such behavioral patterns can lead to pertinent questions regarding the monitoring procedures not recorded in the data
base.

With analytical and sampling "outliers" deleted from the data
base the search for additional patterns was undertaken. Figure 6
shows the multidimensional behavior that characterizes WX, WY, WA, WS
waters. (Note that several monitoring wells were recompleted during
the monitoring program. They are depicted with open squares on this
plot.) The shaded regions depict the multidimensional characteristics
that permit qualitative generalizations. Surface (WS) and alluvial
(WA) waters overlap and exhibit similar chemistries as expected. The
deep upper aquifer well water's (WX's) variance is concentrated along
the vertical axis, while the deep lower aquifers exhibit variance
along both axes. An example of the validity of these generalizations
may be illustrated by examining Figure 7 which shows only those
samples designated as WY type.The three points located at the lower
left portion of Figure 7 have been confirmed as WX samples incorrectly labeled in the field as WY's. Additional information regarding

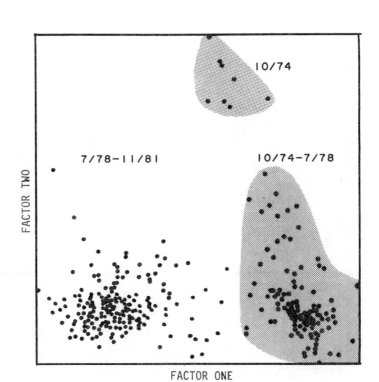

Figure 4. Factor two (B, Cu, Fe, Mn, Ni, Na) vs. Factor one (As, Ba, COD, Li, pH, Sr) factor score plot for WA and WS wells only. Analytical bias identified.

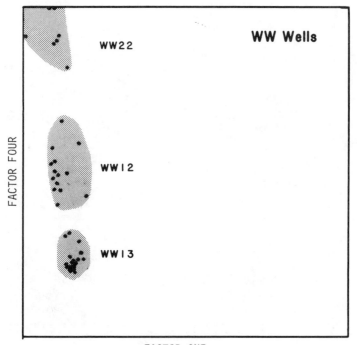

Figure 5. Factor four (NH3, KjN) vs. Factor one (salinity) factor
score plot for WW wells only. Anomalous sample sites identified.

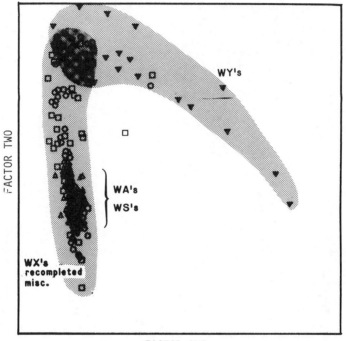

FACTOR ONE

Figure 6. Final Factor two (hardness) vs. Factor one (salinity) factor score plot for 364 samples: WX (○), WY (▼), WA (△), WS(●), R (□) wells.

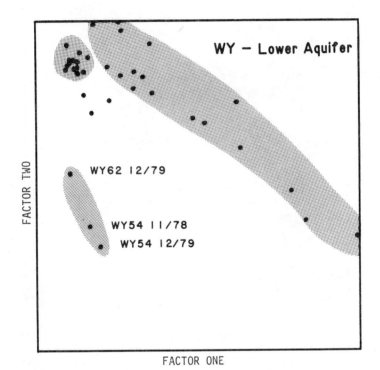

Figure 7. Final Factor two (hardness) vs Factor one (salinity) factor score plot for WY wells only. Characteristic WY behavior is shown. Three WX samples mislabeled as WY are shown.

specific aquifer identity was obtained by examining "fine structure" in the individual category plots. Figure 8 shows the WX waters only. The shaded regions of the plot depict distinctive behavior among upper aquifer waters. These waters have been identified as originating in separate aquifers within the upper aquifer zone. They are easily identfied by the distinctive behavior of the chemical species that contribute to separations along the vertical axis. The existence of separate chemical behaviors indicates that there are "chemical fingerprints" that may be used to identify the subsurface origin of the monitored waters.

A crucial step in examining water quality data is identifying the underlying chemical relationships that determine the complex behavior depicted in these plots. Factor analysis algorithms are used to generate the best composite axes to reproduce the total data variance. As stated earlier, these axes can be visualized as linear combinations of the original measured parameters. The relative contribution made by each variable indicates the importance of each variable in explaining the total variance of the data base. Interpreting the factors provides valuable insight for understanding the water chemistries.

Two factors characterized most of the waters sampled in the monitoring program. The factor loadings for Factor one indicate that the following chemical species participate in correlated behavior that permits the separations and distinctions described above: alkalinity, bicarbonate, B, Cl, conductance, F, Li, Mo, and Na. To simplify discussions in the plots shown earlier this group of species was called the salinity factor. Specific conductance in natural waters usually correlates well with hardness and not as well with bicarbonate, but the current study finds specific conductance closely related to bicarbonate and unrelated to hardness (Ca, Mg, sulfate). This indicates that the ions responsible for increased conductance are probably not calcium or magnesium, rather they are more likely sodium, fluoride, and chloride.

The second important factor, called the hardness factor for simplicity, includes contributions from Ba, F, hardness, Mg, TDS, Sr, and sulfate. This factor characterizes the upper aquifer waters. One may rationalize the distinction between upper and lower aquifers by hypothesizing a natural "softening" in the lower aquifer where ion-exchange of calcium, magnesium and sulfate occurred with sodium and fluoride. It is interesting to note that fluoride occurs in both factors and it alone provides a good aquifer identifier.

Alluvial well waters and springs are chemically similar. They all exhibit moderate hardness and low salinity. These charactersitics may describe varying degrees of saturation in the uppermost stratum. This study also indicates that the measured water quality parameters are not capable of separating alluvial waters from springs and seeps. Additional parameters are necessary to differentiate the two water types.

Conclusions

Factor analysis techniques and the power of their graphical representation permit rapid identification of anomalous behavior in multidimensional water quality data. In addition, the techniques permit qualitative class distinctions among waters with different geologic

Figure 8. Final Factor two (hardness) vs Factor one (salinity) factor score plot for WX wells only. One WY sample mislabeled as WX is shown.

origins because the waters bear different chemical "fingerprints". By examining the factor loadings one gains valuable chemical insights regarding the underlying chemical equilibria that characterize the aqueous media. While factor analytic data examination is fundamentally context free, i.e., it does not depend on any a priori assumptions regarding hypothesized chemical mechanisms, its success is strongly dependent upon the analyst's knowledge of the data base and the system under study. The iterative process of plot selection, anomaly identification and factor interpretation must be viewed as a dynamic insight enhancement protocol. The technique does not answer questions, it merely focuses attention on the key features that require explanation.

Literature Cited

1. Albano, C.; Dunn, W.; III; Edlund, U.; Johansson, E.; Norden, B.; Sjostrom, M.; Wold, S. Anal. Chim. Acta 1978, 103, 429-443.
2. Chu, K.C. Anal. Chem. 1974, 46, 1181-1187.
3. Clark, H.A.; Jurs, P.C.; Anal. Chem. 1975, 47, 374-378.
4. Cooley, W.W. and Lohnes,P.R.; "Multivariate Data Analysis"; John Wiley & Sons, N.Y., 1971.
5. Davis, J.C.; "Statistics and Data Analysis in Geology"; John Wiley & Sons, N.Y., 1973.
6. Erickson, G.A.; Jochum, C.; Gerlach, R.O.; Kowalski, B.R. Paper #99, 65th Ann. Mtng. Amer. Assoc. Cereal. 1980.
7. Gorsuch, R.L. "Factor Analysis" W.B. Saunders Co., Phil.; 1974.
8. Harper, A.M.; Duewer, D.L.; Kowalski, B.R.; Fasching, J.L.;in "Chemometrics: Theory and Applications" ACS Symposium Series, No.52,pp14-52. B.R. Kowalski, ed.
9. Isenhour, T.L.; Kowalski, B.R.; Jurs, P.C. Crit. Rev. Anal. Chem. 1974, 4, 1.
10. Jellum, E.; Bjornson, I.; Nesbakken, R.; Johansson, E.; Wold, S. J. Chrom. 1981, 147: 221-227.
11. Kowalski, B.R.,ed. "Chemometrics: Theory and Applications" ACS Symp. Ser. No. 52, Amer. Chem. Soc., Wash. D.C., 1977.
12. Kowalski, B.R.; Schatzki, T.F.; Stross, F.H.; Anal. Chem, 1972, 44:, 2176-2180.
13. MacDonald, J.C. Amer. Lab. 1978, Feb., 78-85.
14. McGill, J.R.; Kowalski, B.R. Appl. Spec. 1977, 31:2, 87-95.
15. Malinowski, E.R.; Howery, D.G. "Factor Analysis in Chemistry", Wiley Interscience, N.Y., 1980.
16. Massart, D.L.; Kaufman, L.; Coomans, D. Anal. Chim. Acta 1980, 122, 347-355.
17. Nie, N.H.; Hull, C.H.; Jenkins, J.G.; Steinbrenner, K.; Bent, D.H. "Statistical Package for the Social Sciences" 2nd ed.; McGraw Hill N.Y., 1975.
18. Varmuza, K.; Anal. Chim. Acta., 1980, 144:,227-240.

RECEIVED June 28, 1985

Exploratory Data Analysis of Rainwater Composition

Richard J. Vong[1], Ildiko E. Frank[2], Robert J. Charlson[1], and Bruce R. Kowalski[2]

[1]Environmental Engineering and Science Program, Department of Civil Engineering, University of Washington, Seattle, WA 98195
[2]Laboratory for Chemometrics, Department of Chemistry, University of Washington, Seattle, WA 98195

While some aspects of rainwater composition are understood, a large number of important questions remain unresolved, particularly those relating to sources and controlling factors. In search for the chemical and meteorological factors controlling rainwater composition we have utilized SIMCA, PLS, principal component factor analysis, and cluster analysis in the analysis of data consisting of rainwater samples collected in Western Washington State in 1982-83. Major steps of this type of analysis include initial data scaling and transformation, outlier detection, determination of the underlying factors, and evaluation of the effect of experimental error. To reduce potential masking of source-receptor relationships by meteorological variability a data normalization technique was utilized. The components identified for Western Washington rainwater were interpreted to represent the influence of atmospheric oxidation of sulfur and nitrogen compounds, seasalt, soil, and the emissions of a nearby copper smelter.

Considerable interest in the composition of rainwater has been expressed by members of the scientific community in the United States and elsewhere. "Acid rain" has been suggested as the culprit for observed degredation of terrestrial and aquatic ecosystems in the Northeastern United States, Canada, Germany, and Scandanavia. While some aspects of rainwater composition are understood, a large number of important questions remain unresolved, particularly those relating to sources and controlling factors.

Studies of rainwater composition typically include the measurement of the concentrations of a number of chemical species, conductivity, and rain volume and sometimes include supporting measurement of winds or other meteorological parameters. Much of the desired

0097-6156/85/0292-0034$06.00/0

information, the intercorrelations among the measurements, may
remain hidden in the complexity of the data. Multivariate pattern
recognition techniques attempt to identify underlying factors contained in the measurements while reducing the dimensionality of the
data. The measurement of available information (such as the concentration of an element) is used as a step towards identifying these
underlying factors since the factors, themselves, are not directly
measureable (e.g., the influence of a smelter or seasalt). In a
search for the chemical and meteorological factors controlling rainwater composition we have demonstrated the performance of these
techniques in the analysis of data consisting of rainwater samples
collected weekly at three sites in Western Washington State in
1982-83.
    The approach we have undertaken involves the identification of
the underlying factors governing precipitation composition at individual sites supplemented by identification of the factors which
link the local composition at different sites within a region.
Major steps in this type of analysis include initial data scaling
and transformation, outlier detection, determination of the
underlying factors, and evaluation of the effect that experimental
procedures may have on the variance of the results. Most of the
calculations were performed with the ARTHUR software package (1).

Methodology

We have combined classical statistical techniques with graphical
techniques which allow the user a more direct interaction with the
data than would be achieved by a "black box" operation of purely
mathematical techniques.
    For a data set where many samples are available the data reduc-
tion begins with treatment of missing values by elimination of
samples with more than one missing measurement to avoid introducing
bias associated with filling out a large number of missing values.
Single missing values are mean-filled. Due to the low concentrations of many species in rain, measurements below the detection
limit of the analytical technique must be specially treated.
Substitution of a random number between zero and the lower detection
limit avoids introducing correlations which would occur if a
constant or zero value is used. This approach preserves the useful
information that the undetected specie has a very small concentration relative to other samples and to other species.
    A problem in the analysis of these data is the potential
masking of some sources of variability by other correlated variables
which may be difficult to quantify. For example, the potential
meteorological influences of atmospheric dispersion and mixing,
scavenging differences between warm and cold clouds, variable rates
of oxidation of sulfur and nitrogen species, and the dilution effect
of variable rain volume may mask source-receptor chemical relationships. A particular problem is that meteorological data and
source-receptor locations share directional dependence.
    To help reduce these influences, various data normalization
techniques may be applied. Analysis of deposition (concentration
times volume) rather than concentration alone may help avoid variability associated with precipitation amount. Another approach which
was previously applied to aerosol measurements in Sweden (2)

involves converting concentrations to the ratio of an individual
specie to the total concentration of all dissolved species. The
data analysis is then performed on these normalized or relative
concentrations. To the degree that an assumption of constant
scavenging efficiency holds (each element is removed from the
atmosphere with equal efficiency) relative concentrations might be
expected to better reflect the influence of a pollution source,
which, over time might experience differing amounts of dilution by
air and water. This technique may produce spurious correlations due
to closure (the constant sum) depending on the data structure before
normalization (3).

The multivariate techniques which reveal underlying factors
such as principal component factor analysis (PCA), soft independent
modeling of class analogy (SIMCA), partial least squares (PLS), and
cluster analysis work optimally if each measurement or parameter is
normally distributed in the measurement space. Frequency histograms
should be calculated to check the normality of the data to be analyzed. Skewed distributions are often observed in atmospheric
studies due to the process of mixing of plumes with ambient air.
They should be transformed before further data analysis (4). Often
the natural logarithm will convert a skewed distribution to a
roughly gaussian shape. All further data analysis is performed on
these transformed measurements. Normalized or transformed
measurements are termed "features" in the following discussion.

Pattern recognition techniques represent each sample as a point
in N-dimensional space, their coordinates along the axes are the
values of the corresponding measurements. For only two measurements
per sample this is equivalent to representing the sample as a point
on standard two dimensional graph paper. Projection of N-dimensional data onto two dimensional principal component plots provides
a good demonstration of the fundamentals of any multivariate
technique. As in two dimensional graphical techniques the data must
be scaled before further analysis. If no a priori knowledge about
the importance of the different features is available, scaling is
done to equally weight the variance of each feature. A common
approach is termed "autoscaling" (5) where the mean of a feature is
subtracted followed by normalization by the total variance of that
feature. In this manner each feature is transformed to a zero mean
and unit variance. Alternatively, the features may be weighted to
reflect the uncertainty in their measurement, thus giving poorly
determined features less influence on the result (6).

SIMCA and PLS techniques generally utilize a training set for
modeling and predicting the underlying factors in the data and for
classification of unknown samples. This training set must be homogeneous and representitive of the data to be modeled and/or
classified. Therefore, once the initial data scaling and transformation is completed it is important to identify outliers among
the samples so that they will not bias the estimation of model parameters. Identification of outliers also aids in identification of
controlling factors when the pecularities of a particular sample can
be explained in terms of physical processes. We have used exploratory data analysis tools to eliminate outlier samples and choose the
most informative features. Cluster analysis and PCA group the data
in the measurement space to observe natural clusters and outliers.
Projection of the samples onto the first two principal component

axes which represent the bulk of the variance identifies outliers as samples far from the rest of the data. Figure 1 is an example of a principal component projection for rainwater samples collected at the Tolt reservoir site near Seattle, Washington, projected onto axes representing seasalt and aerosol principal components. One sample near the upper left corner of the plot is far from the bulk of the data and is considered an outlier.

The determination of the underlying factors which affect the precipitation composition at a site is done by PC analysis in combination with clustering of sample features. The first step in this process is to identify the "intrinsic dimensionality", the number of controlling factors which are significant in characterizing the rainwater composition. The original number of features are thus reduced to a smaller number of components which contain the information of those original features. The choice of significant factors for a site can be verified by cross-validation (7).

The determination of which features the underlying factors are composed of provides a basis for attaching a physical interpretation to the factors. Varimax rotation of the PCA may be utilized to aid in the interpretation of the factors. Hierarchical dendrograms indicate feature clusters whose composition are analogous to PC factors. The physical interpretation of the clusters and principal components indicates the influence of pollution emission sources or meteorological processes on the rainwater composition at an individual monitoring site.

If the original data contain information on the uncertainties associated with each measurement the sensitivity of the variance of the results to these errors can be studied. Approaches include uncertainty weighting during the autoscaling procedure which is provided for in ARTHUR, uncertainty scaling (the data standard deviation used for autoscaling is replaced by the measurement absolute error such as presented in Table VII), and Monte Carlo simulation for estimating the variance of the statistics based on the error perturbed data (6).

After determining the underlying factors which affect local precipitation composition at an individual site, an analysis of the similiarity of factors between different sites can provide valuable information about the regional character of precipitation and its sources of variability over that spatial scale. SIMCA (8) is a classification method that performs principal component analysis for individual classes (sites) and then classifies samples by calculating the distance from each sample to the PCA model that describes the precipitation character at each site. A score of percent samples which are correctly classified by the PCA models provides an indication of the separability of the data by sites and, therefore, the uniqueness of the precipitation at a site as modeled by PCA.

Spatial interrelationships in the chemical composition among two or more blocks (sites) can be calculated by partial least squares (PLS) (9). PLS calculates latent variables similiar to PC factors except that the PLS latent variables describe the correlated (variance common to both sites) variance of features between sites. Regional influences on rainwater composition are thus identified from the composition of latent variables extracted from the measurements made at several sites. Comparison of the results

obtained from PCA, SIMCA, and PLS models allows the data analyst to
separate local and regional influences on precipitation composition.

Results

We have applied the above approach to a data base consisting of
weekly measurements of 14 chemical species in Western Washington
State rainwater (ammonium, nitrate, chloride, sulfate, arsenic,
cadmium, copper, lead, zinc, potassium, magnesium, sodium, calcium,
and hydrogen ion from pH), conductivity, rainfall volume, rainfall
rate, surface wind speed (U), and frequency of wind direction from
four sectors (NE, SE, SW, NW). Samples were collected at three
sites, in Seattle and in the foothills of the Cascade Mountains in
Washington State over one year (10). Figure 2 indicates the
location of the monitoring sites and a nearby copper smelter which
is a major sulfur dioxide emission source. Additional emissions
occur in the Seattle area, primarily between the West Seattle and
Maple Leaf sites. The wind rose presents data for the frequency
that the wind is from a given direction. Variation in composition
associated with wind direction was deliberately minimized in advance
by site selection directly downwind of the smelter.
    The chemical analyses were performed in the USEPA Manchester,
WA water quality labs by atomic absorption and autoanalyzer
techniques. Charge balance calculations indicated that all
dissolved species of significance were analyzed. Comparison of
filtered and unfiltered aliquots suggested that un-ionized species
were not present in appreciable quantities. Sampling and analysis
uncertainties were determined by the operation of two co-located
samplers for 16 weeks. The calcium and sulfate data were corrected
for the influence of sea salt to aid in the separation of the
factors. This correction was calculated from bulk sea water
composition and the chloride concentration in rainwater (11). Non
seasalt sulfate and calcium are termed "excess" and flagged by a *
in the following discussion.
    Histograms revealed approximately lognormal distributions for
Cl, Na, Mg, K, Ca*, As, Pb, Cd, Cu, Zn and H so those features were
transformed by the natural logarithm. SO_4, NO_3 and NH_4 distributions were roughly gaussian and were not transformed.
    After initial data reduction (treatment of missing values,
transformation and autoscaling) cluster analysis and PCA were used
to visually identify outliers among the samples and to determine
which features did not contribute to the interpretation of the
underlying factors. PCA and cluster analysis were performed first
on the transformed and scaled but unnormalized data. Figure 3
presents the dendrogram (complete link method) for the clustering of
all 22 chemical concentrations and meteorological features at the
West Seattle site. Variables connected at high similarity values on
this dendrogram contain similar information about the rainwater
composition. Relatively tight groupings exist for Na, Mg, and Cl or
for NO_3 and NH_4. The separate branch for As, Pb, Cu, SO_4^*, H^+, wind
speed, SW wind direction, and Cd demonstrates that these variables
are connected with the remainder of the data set at very low
similarity values. This is consistent with a separate source of
variability in the data due to emissions from the Tacoma copper
smelter (the smelter routinely reduces emissions during low wind

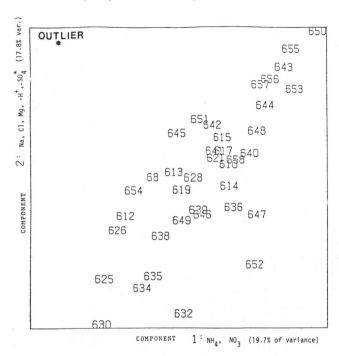

Figure 1: Principal component projection for rainwater samples collected at the Tolt River site.

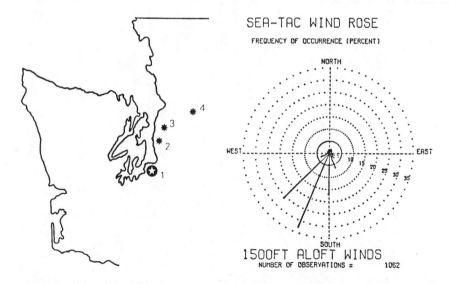

Figure 2: Map of Western Washington with wind direction during rain and locations of the Tacoma copper smelter (1) and monitoring sites at West Seattle (2), Maple Leaf (3), and the Tolt Reservoir (4).

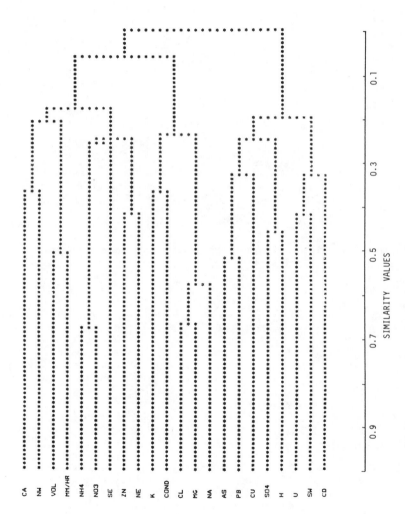

Figure 3: Hierarchical dendrogram for the clustering of all 22 features (unnormalized) at the West Seattle site.

speed or winds from the north). Dendrograms for the other sites were similar. Zn, K, conductivity, and some of the meteorological data were subsequently eliminated from the data set because they did not contribute to the separation of factors or the interpretation of the results. PC projections (such as illustrated in Figure 1) and clustering of samples (as opposed to clustering of features which is displayed in Figure 3) were used to identify rainwater samples which were outliers (as previously described) and might bias the estimation of the PLS and PCA model parameters. These samples were eliminated from the data set before further analysis.

Principal component factor analysis followed by varimax rotation of six factors was performed on four different subsets of the remaining data (each with different preprocessing):
1) Concentration of 14 species with wind direction and rainfall amount,
2) Concentration of 12 species,
3) Deposition of 12 species (concentration times rainfall amount),
4) Fractional concentration of 12 species.

The results of the PCA from each subset are similar except that the data subsets which did not either include the meteorological data or normalize the data to reduce meteorological variability (subsets 2 and 3) were not able to separate several of the components probably due to the atmospheric masking effect. Information on the wind direction and rainfall quantity dependence of seasalt and metals is obtained when meteorological data are included in the analysis. From the standpoint of separation of chemical factors the fourth subset (normalization to fractional composition) provided the best resolution of the data. Using deposition or concentrations, a component that indicated a combined influence of sulfate, nitrate, lead and calcium emission sources was resolved into separate components when the fractional composition data were analyzed by PCA.

In the interpretation of these results it is important to consider the normalization of the data to fractional concentrations and potential spurious correlations due to closure of the data set. Recent work (3) indicates that closure is not a problem when the data set consists of more than eight variables of equal means and variance. If one or several variables are large relative to the others, closure may result in an artificial negative correlation between the larger variables and, sometimes, a positive correlation among the smaller variables. Comparison of the pairwise correlations from the rainwater concentrations to the correlations for the normalized concentrations for our data reveals that only the hydrogen and sulfate correlations with sea salt elements are appreciably altered when the data set is closed. These elements are large relative to the rest of the data (SO_4^* is approximately 40 percent of the total ionic mass) such that closure might be influencing the negative correlation between seasalt elements and SO_4^*/H^*. However, physical processes present an alternate explanation which indicates that this negative correlation would be expected to actually occur as follows: 1) seasalt (Na, Cl, Mg) should be a higher fraction of the ions in winter when high wind speeds generate more salt particles, 2) hydrogen ion and SO_4^* should be a higher fraction of the ions in summer when low wind speeds produce less atmospheric dispersion. When the data are not

normalized all ions were more concentrated during summer when
rainfall volume was small.

Apparently meteorology dominates the fluctuations in
composition in such a manner that the separate pollution influences
could be observed only after meteorological variability, especially
variable rainfall volume, was reduced by the normalization procedure. Since the normalization technique helps to reduce variability associated with atmospheric dispersion and scavenging, this
result implies that meteorological variability was an important
influence on these data.

The weekly sampling period resulted in a variety of meterolog-
ical conditions for each sample and, therefore, precluded any
resolution of samples by unique wind direction or representative
rainfall rate. Therefore, it was not possible to directly evaluate
these meteorological influences on the composition of our
precipitation samples.

Tables I, II and III present the results of the PCA for the
three sampling sites for the fractional concentrations of 12
species. All loadings greater than 0.3 are included. These data
indicate separate influences of Na, Mg, Cl (interpreted to represent
seasalt), NH_4^*, NO_3, SO_4, and H^+ (acid aerosol) and As, Cd, and Pb
(smelter marker elements). The exact combinations of these species
vary from site to site. Hydrogen ion was associated with sulfate at
the West Seattle site but with both sulfate and nitrate at the other
two sites. This is in agreement with the location of major SO_2 and
NO_x emission sources. An additional factor involving Pb and Ca^*
was observed at two sites. This is interpreted to represent the
influence of local soil or road dust. These results account for
about 87 to 91 percent of the total variance in the original data
set. The possible spurious negative correlations between seasalt
elements and sulfate are flagged to note the possible influence of
closure.

Since the PCA and cluster analysis results were similar for the
three sites and since one emission source has been suggested (12) as
the source of many of the species detected in Western Washington
rain, an analysis of the regional similarities in composition was
appropriate.

SIMCA modeling was utilized to determine the separability of
the samples collected at the three different sites. The results
presented in Table IV indicate the model cannot separate the samples
from the West Seattle and Maple Leaf sites. Since both of these
sites are located downwind of the major regional emission sources
and experience similar meteorology their rainwater composition is
similar. The Tolt reservoir site is separated from the Seattle
sites with 79 percent of the samples collected there correctly
classified by the SIMCA model. This site is believed to be
influenced by the same emission sources as the other two sites but
experiences different meteorological conditions (primarily longer
transport times and more frequent and larger quantity of rainfall)
due to its location in the foothills of the Cascade Mountains
(elevation 550 meters). Considering the uncertainty in the reported
concentrations (see Table VII) and the similar air pollution
emission sources the SIMCA results are reasonable.

The final step in the analysis was utilization of PLS to examine the correlated variance of the features between different sites.

Table I: West Seattle (fractional concentrations)

Varimax Rotation of Principal Factor Pattern

Specie	a_1	a_2	a_3	a_4	a_5	a_6
NH_4		.610				
NO_3		.625				
Cl	-.460**					
SO_4*	.573**				.486	
As				.716		
Cd				.570		
Cu						.925
Pb			.464	.468		
Na	-.455**					
Mg	-.469**					
Ca*			.731			
H					.890	
Percent of Total Variance	19.3	18.8	13.8	12.6	9.5	8.8

*Corrected for seasalt based on chlorinity ratio.

**These loadings are believed to be realistic although a potential closure problem exists (see text).

Table II: Maple Leaf (fractional concentrations)

Varimax Rotation of principal factor pattern

Specie	a_1	a_2	a_3	a_4	a_5	a_6
NH_4		.634				
NO_3		.408	-.556			
Cl	-.481**					
SO_4^*	.534**					
As						.320
Cd						.910
Cu				.834		
Pb					.776	.373
Na	-.450**					
Mg	-.302**					
Ca^*				.369	.413	
H	.433**		-.433			
Percent of Total Variance	26.7	15.9	14.9	10.9	10.4	8.8

*Corrected for seasalt based on chlorinity ratio.

**These loadings are believed to be realistic although a potential closure problem exists (see text).

Table III: Tolt Reservoir (fractional concentrations)

Varimax Rotation of Principal factor patterns

Specie	a_1	a_2	a_3	a_4	a_5	a_6
			Factor Loadings			
NH_4	.607					
NO_3	.568			.365		
Cl		.392**		-.622		
SO_4^*		-.602**		.363		
As					.930	
Cd				.688	.332	
Cu			.624			
Pb				.605	.355	
Na		.456**				
Mg		.533**				
Ca^*			.663			
H		-.378**			.500	
Percent of Total Variance	19.7	17.8	13.7	13.3	12.1	9.1

*Corrected for seasalt based on chlorinity ratio.

**These loadings are believed to be realistic although a potential closure problem exists (see text).

Table IV: SIMCA results, classification matrix for fractional
concentrations at three sites.

		West Seattle	Maple Leaf	Tolt River
	West Seattle	19 51% correct	7	11
True Class	Maple Leaf	20	23 45% correct	8
	Tolt	4	3	26 79% correct

Any regional influence on rainwater composition would be expected to affect all three sites reported here. A PLS two block model (9) was used to predict the variance in rainwater composition at one site from the variance in rainwater composition at an upwind site.

PLS results are presented in Tables V and VI. Loadings greater than 0.3 have been underlined. Loadings which may be influenced by closure are flagged. The regression of Maple Leaf composition (fractional) on West Seattle composition reveals four components:

la) Hydrogen ion, lead, sulfate, nitrate (positive correlation);
lb) Sodium (negative correlation);
2a) Arsenic, cadmium, lead (positive correlation);
2b) Nitrate (negative correlation);
3) Sodium, magnesium, chloride, ammonium;
4) Cadmium, copper.

The first three components suggest regional sources of: acidic anthropogenic aerosol, the marker elements of a copper smelter, and seasalt, respectively. The fourth component or the ammonium in component three do not provide a ready interpretation of a known emission or meteorological source of variability. The negative correlation of nitrate with component two is consistent with separate influences of the copper smelter and automobile emissions.

The regression of the Tolt River rainwater composition on Maple Leaf data indicated four components:

la) Sodium, magnesium, chloride (negative correlation);
lb) Hydrogen ion, sulfate lead (positive correlation);
2) Arsenic, cadmium, lead;
3) Copper, lead
4a) Sulfate, magnesium (negative correlation)
4b) Ammonium (positive correlation)

Three components are similar to the results for the West Seattle—Maple Leaf PLS model except that the acid aerosol component no longer has high a loading from nitrate. This specie is ordinarily associated with automobile emissions. The Tolt site is remote enough that auto emissions are not as important an influence on the variability in rainwater composition as in Seattle. The fourth component for this PLS model might represent emissions from a cement plant which does not influence the West Seattle site. The soil factor is apparently local in nature since it appears in the PCA results but not the PLS results.

With emission source chemical signatures and corresponding aerosol or rainwater sample measurements PLS can be used to calculate a chemical element mass balance (CEB). Exact emission profiles for the copper smelter and for a power plant located further upwind were not available for calculation of source contributions to Western Washington rainwater composition. This type of calculation is more difficult for rainwater than for aerosol samples due to atmospheric gas to particle conversion of sulfur and nitrogen species and due to variations in scavenging efficiencies among species. Gatz (14) has applied the CEB to rainwater samples and discussed the effect of variable solubility on the evaluation of the soil or road dust factor.

Table VII presents data for Maple Leaf rainwater collected in two co-located samplers operated for 16 weeks for the purpose of determining experimental uncertainty. These data reveal that Cu,

Table V: Outer relationship coefficients of the PLS model

Two block PLS: West Seattle ——> Maple Leaf

Specie	NH_4	NO_3	Cl	SO_4^*	As	Cd	Cu	Pb	Na	Mg	Ca*	H
1. latent variable												
West Seattle	.31	.39	-.37**	.22**	.22	.21	.03	.37	-.39**	-.26	.09	.34**
Maple Leaf	.16	.33	-.05**	.43**	.16	.23	.09	.40	-.17**	.04	.31	.39**
2. latent variable												
West Seattle	.23	.37	-.21	-.04	-.57	-.30	-.25	-.38	.08	-.34	-.15	.01
Maple Leaf	.16	.33	.13	.01	-.51	-.45	-.15	-.39	.20	-.07	-.41	.10
3. latent variable												
West Seattle	.32	.23	.41	-.33	.23	.06	-.23	.29	.30	.30	-.24	.10
Maple Leaf	.39	.28	.44	.19	.29	.06	-.15	.24	.36	.37	.13	.01
4. latent variable												
West Seattle	-.04	-.03	-.14	.08	-.21	.73	-.56	-.11	.01	.16	.15	-.13
Maple Leaf	.08	.13	-.17	.15	-.29	.80	-.42	-.08	.06	.06	.11	.12

* Corrected for seasalt based on chlorinity ratio.
** These loadings are believed to be realistic although a potential closure problem exists (see text).

Table VI: Outer relationship coefficients of the PLS Model

Two block PLS: Maple Leaf ——> Tolt Reservoir

Specie	NH$_4$	NO$_3$	Cl	SO$_4$*	As	Cd	Cu	Pb	Na	Mg	Ca*	H
1. latent variable												
Maple Leaf	.15	.29	-.48**	.31**	-.18	.09	.01	.32	-.35**	-.38**	.20	.36**
Tolt reservoir	.12	.18	-.43**	.28**	-.07	-.04	.10	.36	-.52**	-.40**	.17	.33**
2. latent variable												
Maple leaf	.03	-.21	-.09	.08	.72	.28	.13	.29	.08	.03	.46	-.14
Tolt reservoir	.03	-.20	.09	.01	.71	.44	.01	.29	.22	.09	.26	-.18
3. latent variable												
Maple leaf	-.33	.10	-.22	.07	.29	-.29	.49	-.55	.08	-.21	.03	.25
Tolt reservoir	.20	.05	.04	.20	.27	-.02	.44	-.63	.12	-.01	.49	.01
4. latent variable												
Maple leaf	.40	.33	.03	-.47	.14	-.25	.22	.26	.28	-.43	-.15	-.17
Tolt reservoir	.39	.16	-.33	-.30	.40	.27	.33	.05	.13	-.47	-.21	-.06

* Corrected for seasalt based on chlorinity ratio.
**These loadings are believed to be realistic although a potential closure problem exists (see text).

Table VII: Sampling and analysis precision for co-located rain
              samplers of the Maple Leaf site (units = ppm unless
              indicated)

Species	Mean (1)	Absolute Error (2)*	(2)/(1)	N
NH_4-N	0.30	0.05	.15	32
NO_3-N	0.45	0.07	.16	32
Cl	1.04	0.16	.15	32
SO_4	3.67	0.81	.22	32
As(ppb)	7.18	1.94	.27	20
Cd(ppb)	0.65	0.74	1.15	26
Cu(ppb)	7.97	5.66	.71	32
Pb(ppb)	17.3	3.73	.22	30
Zn(ppb)	17.5	7.14	.41	26
K	0.13	0.08	.66	32
Na	0.68	0.08	.12	30
Mg	0.10	0.02	.15	30
Ca	0.26	0.04	.16	30

*The standard deviation was calculated assuming that the average of
each co-located sample pair was the pair's true value. Random
error was assumed. N/2 degrees of freedom were used for the N/2
sample pairs since no overall mean for the data set was calculated.
The absolute error is defined as this standard deviation of paired
sample collections for a 16 week period. The data for the entire
52 week sampling period have been reported elsewhere (10).

Cd, K and Zn are not precisely determined. Previously reported (13) results for identical split samples indicates that most of this experimental error was due to analytical imprecision rather than collection and handling. Many of the samples were near the detection limit for the five trace metals (As, Cd, Cu, Pb, Zn). To determine the effect of these measurement errors the PCA was repeated with uncertainty scaled data. (The data standard deviation used in autoscaling was replaced with the measurement absolute error.)

The effect of including the measured analytical and sampling errors in the data scaling and PCA was to split factors consisting of several trace metals (which had higher uncertainties than the other species). In many cases the error weighted PCA indicate primarily single features such as arsenic, cadmium, or copper loading on a component. This is consistent with a source of variance in the data set which is associated with random measurement variations rather than emission sources or meteorological processes. This emphasizes the importance of using accurate and precise analytical techniques for rainwater measurements.

Conclusions

The four techniques (PCA, hierarchial clustering, SIMCA, and PLS) are complementary in resolving precipitation chemistry data. Interpretation of these results allows a hypothesis as to what factors influence precipitation chemistry in Western Washington. Since the choice of which species to chemically analyze is subjective, other factors may be undetected due to lack of measurement. These results indicate the presence of seasalt, acidic sulfate and nitrate aerosol, road or soil dust, emission of metals from a copper smelter located to the southwest, and the occurrence of rain accompanied by strong southwesterly winds. These results are consistent with previous work (15). Further identification of meteorological influences on composition is limited by the weekly sampling period which results in a variety of wind and rain patterns for each sampling period.

Although the measurement uncertainties limit the conclusions which can be drawn from these results, the data set proved useful for the determination of general influences on rainwater composition in the Seattle area and for the demonstration of the application of these exploratory data analysis techniques. Current efforts to collect and analyze aerosol and rainwater samples over meteorologically appropriate time scales with precise analytical techniques are expected to provide better resolution of the factors controlling the composition of rainwater.

Literature Cited

1. Duewer, D.L., Harper, A.M., Koskinen, J.R., Fasching, J.L., and Kowalski, B.R., ARTHUR, Version 3-7-77 (1977).
2. Hansson, H.C., Martinsson, B.G., and Lannefors, H.O., accepted for publication in Nuclear Instruments and Methods, (1984).
3. Johansson, E., Wold, S. and Sjodin, K, Analytical Chemistry, 56, 1685, (1984).

4. Brown, S.D., Skogerboe, R.K., and Kowalski, B.R., Chemosphere, 9, 265, (1980).
5. Kowalski, B.R. and Bender, C.F., J. Am. Chem. Soc., 94, 5632 (1972).
6. Duewer, D.L., Kowalski, B.R., and Fasching, J.L., Anal. Chem., 48, 13, 2002, (1976).
7. Wold, S., Technometrics, 20, 4, 397, (1978).
8. Wold, S., J. Pattern Recognition, 8, 127, (1976).
9. Frank, I.E. and Kowalski, B.R., J. Chem. Inf. Comput. Sci., 24, 1, 20, (1984).
10. Vong, R.J., Larson, T.V., Covert, D.C., and Waggoner, A.P., accepted for publication Water, Air, and Soil Pollution, (1985).
11. Junge, C., Air Chemistry and Radioactivity, New York (1963).
12. Larson, T.V., Charlson, R.J., Kundson, G.J., Christian, G.D., and Harrison, H., Water, Air, and Soil Pollution, 4, 319, (1975).
13. Vong, R.J. and Waggoner, A.P., EPA 910/9-83-105, USEPA Region 10, Seattle, WA. (1983).
14. Gatz, D.F., "Source Apportionment of Rain Water Impurities in Central Illinois," presented at 76th A.P.C.A. Meeting, Atlanta, (1983).
15. Knudson, E.J., Duewer, D.L., Christian, G.L. and Larson, T.V. in: Chemometrics, Theory and Applications, (ed. B.R. Kowalski), ACS Symposium Series 52, Wash, D.C. (1977).

RECEIVED June 28, 1985

Multivariate Analysis of Electron Microprobe–Energy Dispersive X-ray Chemical Element Spectra for Quantitative Mineralogical Analysis of Oil Shales

Lawrence E. Wangen, Eugene J. Peterson, William B. Hutchinson, and Leonard S. Levinson

Chemistry Division, Los Alamos National Laboratory, Los Alamos, NM 87545

Methods for determining the mineral content of complex
environmental samples by multivariate analysis of
chemical element data are under investigation. Major
elements are determined by simultaneous analysis of the
energy dispersive spectra from an electron microprobe
system. Elemental data and size are obtained for 1000
locations on a single shale sample. The elemental data
are analyzed by clustering methods to determine inherent
sample groups and to produce sample subsets containing
fewer mineral components. These subsets are analyzed by
target transformation factor analysis to determine (1)
the number of significant mineral components; (2) the
physically meaningful mineral component vectors; (3) the
contributions of each mineral component to the elemental
concentrations of each sample location; (4) the quantity
of each mineral component at each sample location; and
(5) the mineral composition of the entire sample. X-ray
diffraction data and x-ray intensities (from energy
dispersive analysis) of elements in pure minerals known
to occur in oil shale aid in interpreting mineral component vectors. An overview of the method will be
presented with results of its application to a raw oil
shale sample.

Quantitative determination of the major and minor minerals in
geological materials is commonly attempted by x-ray diffraction
(XRD) techniques. Mineralogists use a variety of sophisticated and
often tedious procedures to obtain semiquantitative estimates of the
minerals in a solid sample. The mineralogist knows that XRD intensities depend on the quantity of each mineral component in the
sample even through expressions for conversion of signal intensity
to quantitative analysis often are unknown or difficult to determine. Serious difficulties caused by variables such as particle
size, crystallinity, and orientation make quantification of many
sample types impractical. Because of a lack of suitable standards,
these difficulties are particularly manifest for clay minerals.
Nevertheless, XRD remains the most generally used method for quan-

0097–6156/85/0292–0053$06.00/0

tifying the mineral components of solid geological materials,
probably because it is the best method for qualitative identification of minerals in complex mixtures.

Recently, methods for quantitatively determining the chemical
element composition of solid materials by x-ray emission spectroscopy using the electron microprobe have become available. A significant advantage of the electron microprobe, compared with methods
for bulk analysis, is its capability for rapid analysis of many
different micron-size areas of a solid sample. Thus, in a relatively short time, we can obtain several hundred quantitative
analyses of the chemical element content of a solid sample. These
analyses usually will be different because sample homogeneity is
absent on the micron level. Thus, each chemical analysis is a
linear sum of the chemical elements in the subset of minerals
present at that location. Generally, we expect the number of
minerals present in a micron-size spot to be less than the total
number of minerals in the bulk sample.

Recent work reported a method for estimating the mineral content of coals based on the electron-microprobe-determined chemical
composition of discrete particles. (1,2) Each particle is assumed
to contain only one mineral component. Possible ambiguities in
qualitative identification of discrete mineral particles can be
eliminated by XRD analyses of the bulk material to identify the
minerals present. For most geological materials, such separations
are not readily obtainable. Thus, this method is limited to materials that can be dispersed into particles composed of single
minerals.

During recent years, multivariate data analysis methods for
determining the number of components in mixtures that are linear
sums of the components have been developed. (3-5) If the number of
components in a mixture can be identified, methods are available for qualitatively identifying them and for determining the composition of each component. These often involve computer searching
of possible compositions to find physically meaningful ones. Alternatively, investigators can guess compositions based on knowledge of
the system under study and determine if these guessed components can
explain the observed data. These multivariate methods are based on
variations of factor analysis and are identified in the literature
by different terms, such as target transformation factor analysis
(TTFA), (3) Q-mode factor analysis, (4) or multicomponent curve
resolution. (3) A recent paper by Roscoe et al. applied this method
to quantitative determination of mineral matter in coals using only
the chemical element concentrations. (6) Their results compared
favorably with results concerning mineral content determined independently by XRD analyses.

The purpose of this paper is to describe procedures under
development for determining the quantitative mineralogical composition of complex geological materials. The approach consists of
the following:
1. quantitative chemical element analysis at several hundred sample
locations on a solid surface by electron microprobe x-ray emission
spectroscopy,
2. assignment of each sample location to one of several clusters
based on chemical element composition using a multivariate data
analysis method called cluster analysis,

3. determination of the number of mineral components in each spot and the composition of each component by TTFA,
4. determination of the fractional contribution of each mineral component to each spot by multiple regression, and
5. determination of bulk mineralogical composition by summation over all spots.

Identification of major and minor components is verified by qualitative XRD analysis or other procedures.

Methods

Experimental Procedure. Oil shale is a fine-grained sedimentary rock that contains an organic material known as kerogen. The Green River formation oil shales underlie approximately 16,000 miles of the tri-state area of Colorado, Utah, and Wyoming. The sample used in this study was obtained from core material recovered from the Piceance Creek basin in northwest Colorado. The mineralogy of this raw shale sample is typical of shales from this region; the XRD determination of the mineralogy is listed in Table I.

Table I.   Qualitative Mineralogical Analysis of Oil Shale by
X-Ray Diffraction

α-Quartz	Strong	Orthoclase	V Weak
Illite	V Weak	Albite	Medium
Dolomite	Medium	Pyrite	Trace
Dawsonite	Weak	Gypsum	Trace
Kaolinite	Trace		

Usually, bulk samples are crushed to less than 10 mm and split to obtain workable quantities of material. Fractions of these are crushed again so the rock passes a 20 mesh sieve and then is ground to -200 mesh. Portions of this material are taken for XRD and electron microprobe energy dispersive x-ray emission (EDX) analysis. Samples for EDX probe analysis are made into 100-mg pellets at 2000 psi. Before analysis, the pellets are coated with 100 to 200 angstroms of carbon.

EDX analysis is accomplished with a Cameca MBX electron microprobe with Tracor Northern automation and an energy dispersive x-ray analyzer. A computer program has been written that will perform up to 1000 analyses for 13 elements in approximately 2 h. The results are stored on floppy disks and can be transferred to main frame computers where multivariate analysis can be performed. The elements monitored during each analysis include sodium, magnesium, aluminum, silicon, phosphorus, sulfur, chlorine, potassium, calcium, titanium, manganese, iron, and copper. Detection limits for the elements analyzed ranged from 0.1 wt% for silicon to 1.0 wt% for copper. Each analysis was run by using 15-keV beam energy, a 5-na beam current, a 0.3-μm beam diameter, and 4-s collection time. The results of each analysis are output as intensity ratios to pure element standards that are first-order approximations of the concen-

tration for a given element. Seventeen mineral standards were analyzed to determine the accuracy and precision of the method. The
accuracy ranged from 10 to 20 rel%, whereas the precision on major
constituents was about 2 rel%. Preliminary work with quantitative
correction procedures has predicted the accuracy to be about 5 rel%.
Future work will include corrections to the intensity ratios to make
the results more quantitative.

Multivariate Data Analysis. After determining the chemical element
composition for 10 to 12 elements of a complex geological sample at
about 1000 locations, we have a 1000 X 10 data matrix. Because we
have been investigating an oil shale sample, each of these 1000
locations can contain kerogen. The amount of kerogen at each location will not usually be the same. Thus, because carbon, hydrogen,
and oxygen are not measured, the element concentrations do not sum
to a constant value for each sample. For this reason, the chemical
element concentrations for each sample location are normalized by
the following procedure:

$$X'_{ik} = X_{ik} / \sum_k X_{ik} \qquad (1)$$

where X_{ik} is the relative concentration from the EDX analysis of
element k in sample i, and the sum is over all determined elements.
After this normalization, the chemical element values for each
sample sum to 1.0.

$$\sum_{k=1}^{NV} x_{ik} = 1.0 \text{ for all } i \qquad (2)$$

where for convenience we drop the prime. The main effect of this
procedure is to remove differences in chemical element intensity
caused by variations in kerogen or organic content.

Cluster analysis on samples is accomplished by using the K
means method incorporated into BMDP. (7) This method finds the
number of clusters requested by the investigator by using the means
centering method. (8) Successively increasing numbers of clusters
can be requested to determine the robustness of clusters. Samples
that remain in the same cluster while increasing numbers of clusters
are formed are thought to be part of a robust cluster.

The K means algorithm in BMDP uses the Euclidian distance as a
measure of similarity between samples. A number of data standardi-
zations are available that give the effect of calculating the distance using original concentrations, mean-centered concentrations,
concentrations standardized by variance, or concentrations mean
centered and standardized by variance. Different methods of standardization were investigated; however, using the raw data with no
standardization gave the best results. This is believed to derive
from the importance of composition ratios such as for calcium and
magnesium in dolomite, iron and sulfur in pyrite, and aluminum,
silicon, and potassium in illite.

In this research, the purpose of the cluster analysis is to
obtain groupings of samples with compositions similar to specific
minerals or with only a subset of all the minerals in the bulk
sample. This is necessary in most geological samples in order to

reduce the mathematical rank of the data matrix to a value smaller
than the number of chemical elements determined. When this is
accomplished, it becomes mathematically possible to determine the
true number of mineral components in each cluster. In contrast, if
a set of samples contains more mineral components than determined
chemical elements, a physically meaningful mathematical solution is
not possible because the rank of a matrix cannot be larger than its
smallest dimension.

After allocating sample locations to specific clusters, each
cluster is subjected to TTFA, a method discussed at length in the
monograph by Malinowsky and Howery (9) and more briefly by Hopke and
colleagues. (6,10-11) The latter have applied the method to source
characterization of air particulate matter and to coal mineralogy.
(6,10) Their computer programs for performing TTFA, FANTASIA, (11)
were used in this research. Most of the following discussion is
based on the work of these researchers.

The basic model of the factor analysis method as applied here
assumes that the x-ray emission intensity of any specified element
is a linear sum of the quantity of that element found in the
minerals present at that sample location:

$$x_{ik} = \sum_{l=1} c_{kl} \, f_{li} \qquad (3)$$

where x_{ik} is the same as in Equation 1, c_{kl} is the average
composition of chemical element k in mineral component l, and f_{li} is
the mass fraction of mineral component l in sample i. As an example
of Equation 3, the amount of aluminum in a sample that is a mixture
of the minerals kaolinite, albite, and gibbsite would be

$$x_{Al} = c_{Al,kao} \, f_{kao} + c_{Al,alb} \, f_{alb} + c_{Al,gib} \, f_{gib} \qquad (4)$$

A similar expression could be written for silicon, potassium, or
sodium. If we have NS = 1000 samples in which NV = 10 chemical
elements have been measured, our model would contain 10,000
expressions, as in Equation 4, one for each element per sample. In
matrix notation, these relations are

$$\underline{X} = \underline{F} \, \underline{C} \qquad (5)$$

where $\underline{C}$ is the NCxNV matrix of average chemical compositions of each
of the NC minerals, $\underline{F}$ is the NSxNC matrix of fractional compositions, and $\underline{X}$ is the NSxNV matrix of measured chemical element
compositions in the NS samples. For our problem, $\underline{X}$ contains the
x-ray emission intensities of the measured elements at each of the
1000 locations on the sample, expressed as relative concentrations.

The objectives of TTFA are
1. to determine NC, the number of mineral components that make up
the set of samples,
2. to identify the average elemental compositions of each mineral
component contributing to the set of samples,
3. to determine the mass fractions of each mineral component in
each sample location, and
4. to sum over all samples and determine the quantitative mineralogical composition of the original solid material.

The first step in the TTFA procedure involves a principal
components factor analysis of the data matrix. This step is accomplished by an eigenanalysis (or characteristic value analysis) of
the product moment of the data ($\underline{X}$) matrix normalized and weighted as
desired by the investigator. No weighting or mean subtraction is
referred to as covariance about the origin, sometimes referred to as
the scatter matrix. Mean subtraction results in a standard covariance matrix. Weighting by the square root of reciprocal variance is
correlation about the origin, and mean subtraction plus this
weighting is standard correlation. To date, most of our work has
used weighting by the square root of reciprocal variance, which is
the method advocated by Hopke and colleagues. ($\underline{6},\underline{10-11}$) This procedure preserves the relative concentration of the chemical element
data expressed in standard deviation units. Other methods of
weighting and data expression might be more appropriate, depending
on the nature of the problem.

Because of error, eigenanalysis of real samples always will
produce NV eigenvalues and associated eigenvectors. In TTFA, we
assume that a subset of the eigenvectors, numbering NC, will account
for all the important data variation not caused by random errors or
unimportant (for present purposes) components. Several tests can be
applied to determine the correct number of eigenvectors. ($\underline{12}$) In
our case, we can use as a realistic upper limit the number of significant mineral phases identified by XRD if such analysis has been
performed. In practice, real samples seldom have a definable number
of mineral phases; rather they have a number of easily identifiable
major and minor minerals, plus numerous trace mineral components
that may or may not be identified. The investigator must use some
judgment in deciding the value of NC. In this work, NC is defined
according to the number of components needed to reproduce the chemical element data for those elements making up the dominant mineral
phases. Such a procedure satisfies our principal purpose of quantifying major and minor mineral components.

After finding NC, we must determine the composition of each
mineral. It is very helpful at this point to have a qualitative
mineralogical analysis, such as XRD, to provide initial estimates of
compositions. In addition, libraries of mineral compositions are
extremely useful. Methods based on searching the original data
matrix for candidate minerals also are helpful and in some instances
may provide the best compositions for real samples.

Candidate mineral compositions or test vectors are tested by
linearly rotating the NC eigenvectors towards a test vector by using
a least squares procedure ($\underline{3}$) and determining if the test vector
could possibly lie in the vector space defined by the NC eigenvectors. In this way, suspected minerals are kept or rejected from
further consideration. (From this step of the analysis, TTFA
derives its name.) When NC mineral compositions have been determined that adequately reproduce the original data and are consistent
with other information, such as XRD or infrared analysis, this
aspect of TTFA is finished. At this point, we have successfully
determined the matrix $\underline{C}$ of Equation 5.

After $\underline{C}$ is determined, $\underline{F}$ is obtained by proper manipulation of
Equation 5. Finally, after proper scaling of the $\underline{A}$ and $\underline{F}$ matrices
and accounting for any sample normalization, such as in Equation 1,

we can sum over all samples to obtain a quantitative estimate of the mineralogical composition of our original solid sample.

Results and Discussion

This work is motivated by a lack of techniques for quantifying the mineral components in complex environmental solids. Programmatic interest derives from research in the environmental chemistry of raw and retorted oil shales from the Piceance Basin in Colorado. For this reason, we chose to do exploratory research on a particular sample of raw shale. Previously XRD analysis had been performed on this sample. The XRD results are shown in Table I. XRD line intensities for the minerals often are used to provide a rough semiquantitative estimate of amount present.

Cluster Analysis. Cluster analysis using BMDP's PKM method was performed on the data with several methods of data transformation, normalization, and variable standardization. Qualitative clustering results for these different procedures of data manipulation were similar. The method finally selected is that discussed above, i.e., normalization of each sample, so the concentrations sum to unity and use of Euclidian distances with no standardization of variables as a measure of sample similarity.

A number of clusters from 4 through 14 were formed successively, each time starting from one cluster containing all samples. The clusters at each level and the progression of cluster formation as new clusters were formed at each successive level through 10 clusters are shown in Figure 1 as a tree diagram. Clusters that maintain their identity as the number of clusters increases are thought to form "robust" clusters. For example, cluster five (18 samples) remained intact, as the number of clusters increased from 5 to 14. At each level, a subset of samples breaks off from one bigger cluster to form a new cluster. The farther down the tree this occurs, the more similar is the new cluster to the one from which it originated. Average normalized elemental concentrations for each cluster at 10 clusters are given in Table II. We chose to proceed with 10 clusters. The XRD results together with the normalized concentrations in Table II indicate the dominant minerals likely to be present in each cluster. Subsequent TTFA analysis was performed separately on each of these 10 clusters.

TTFA. The TTFA steps will be illustrated for the first two clusters in Figure 1. These correspond to clusters one and two in Table II. As evident in Figure 1, these two clusters are quite similar. Based on the relative elemental concentrations in Table II, we refer to these as the silicon and albite clusters.

Real geological samples rarely exhibit a specific number of mineral components. Such samples are composed of a multitude of minerals that are qualitatively referred to as major, minor, and trace components. The sensitivity of the particular analytical method limits our ability to resolve these minerals. Here we determine the number of mineral components (NC) from the eigenanalysis of the raw data matrix. The raw data matrix is approximated by a successively increasing number of eigenvectors. When this approximated data matrix is within the expected uncertainty of the data, we

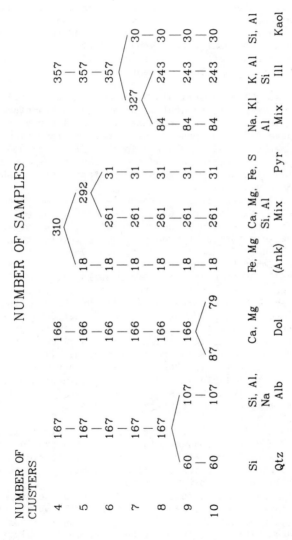

Figure 1.　Tree diagram allocating samples to clusters as number of clusters increases.

Table II. Relative Concentrations in Each Cluster Normalized to a Sum of 1.0

Cluster	Samples	Na	Mg	Al	Si	S	Cl	K	Ca	Ti	Fe
1	60	0.0000	0.0046	0.0144	0.9206	0.0085	0.0002	0.0084	0.0407	0.0012	0.0015
2	107	0.0176	0.0104	0.0813	0.7293	0.0228	0.0012	0.0342	0.0853	0.0017	0.0153
3	87	0.0000	0.1864	0.0181	0.1237	0.0215	0.0003	0.0077	0.6040	0.0001	0.0377
4	79	0.0002	0.1083	0.0546	0.2686	0.0409	0.0000	0.0250	0.4523	0.0023	0.0469
5	18	0.0006	0.1035	0.0335	0.1632	0.0501	0.0000	0.0135	0.0878	0.0019	0.5443
6	261	0.0005	0.0563	0.0845	0.4311	0.0458	0.0002	0.0413	0.2977	0.0044	0.0350
7	31	0.0029	0.0421	0.1074	0.2618	0.1400	0.0000	0.0245	0.2250	0.0251	0.1702
8	84	0.0064	0.0136	0.1529	0.4805	0.0488	0.0179	0.0807	0.1380	0.0052	0.0556
9	243	0.0024	0.0061	0.1360	0.6125	0.0229	0.0006	0.1149	0.0869	0.0016	0.0142
10	30	0.0182	0.0115	0.3806	0.4432	0.0670	0.0000	0.0177	0.0465	0.0006	0.0146
Mean		0.0042	0.0476	0.0994	0.4885	0.0374	0.0027	0.0518	0.2219	0.0032	0.0947

have an estimate of the number of mineral components making up the
samples. Various measures for estimating NC are discussed in
Reference 12. In practice, the value of NC is ambiguous and several
values can be tested; the investigator must use scientific judgment.
At this stage of the research, we are using an average error of 10%
for the reproduced elemental concentrations as an initial guess for
estimating the maximum number of components in each cluster and then
testing candidate mineral vectors, assuming several values of NC
between two or three and the number required to give an average
error of 10%. Table III for the albite cluster (cluster two) shows
FANTASIA's summary printout, which is used to estimate NC. (11)
None of the indicators in the table provide a definitive value for
NC. The eigenvalue shows the amount of total variance explained and
so is heavily weighted in favor of elements present in the samples
at high concentrations, i.e., silicon for this case. Thus, if
concentrations of all elements are to be successfully approximated,
use of the eigenvalue as a measure of the number of components is no
good. Using the 10% average error criterion, we arrive at an NC of
four for the albite cluster. Estimates of the number of mineral
components in the other nine clusters were determined by a similar
rationale.

We have obtained EDX relative concentrations for numerous
naturally occurring "pure" minerals. These are used as test vectors
in target transformation to determine if any of the eigenvectors can
be linearly rotated to fit a test vector. In addition, these "pure"
minerals are used to aid in searching the data matrix for compositions that most closely approximate them. Such elemental concentrations of minerals from the geological sample may be a better estimate of those making up the sample than are mineral standards.
Several elemental concentrations of such mineral standards are
listed in Table IV. These vectors are normalized to a sum of unity
to be consistent with the normalization of the data samples. A
least squares minimization procedure is used to determine the linear
transformation (rotation) that will best transform one of the eigenvectors to the test vector. (13) We compare the new test vector
predicted by this transformation with the original to evaluate
whether the original candidate mineral is one of the components.
This process is repeated for all candidate test vectors at any
desired value of NC. In this way, different numbers of mineral
components can be investigated to determine a set of mineral components that reproduces adequately the original data.

The silicon cluster contained 60 samples and a normalized
concentration of 92% silicon. This cluster was estimated to be
entirely quartz. Only the quartz test vector provided an adequate
fit upon rotation of one or two eigenvectors.

A more complex example is provided by cluster two, the albite
cluster containing mainly sodium, aluminum, and silicon. Consideration of the number of vectors from the eigenanalysis in Table III
suggested the presence of four components. Target testing identified these as albite, quartz, orthoclase, and perhaps gypsum. The
fit of these four vectors is shown in Table V. Normalized elemental
concentrations less than about 0.01 are considered unreliable
because they are too close to the noise level. Thus, for example,
vector one, identified as albite, contains only sodium, aluminum,
and silicon; vector two, identified as quartz, appears to have a

Table III. Principal Components Analysis of Cluster 2 (Albite Cluster)

Factor	Eigenvalue	RMS	Chi-Square	Exner	Average Error
1	1.1e+2	9.9e-4	1.2e-3	0.15	29.2
2	1.2e+0	6.2e-4	5.3e-4	0.09	17.7
3	3.2e-1	4.8e-4	3.6e-4	0.07	15.8
4	2.3e-1	3.4e-4	2.2e-4	0.05	11.1
5	9.7e-2	2.7e-4	1.6e-4	0.04	7.2
6	6.7e-2	2.0e-4	1.1e-4	0.03	4.8
7	3.8e-2	1.4e-4	7.9e-5	0.02	3.0
8	1.7e-2	1.1e-4	7.1e-5	0.02	1.2
9	1.7e-2	6.8e-5	5.4e-5	0.01	0.5

Table IV. Normalized EM–EDX Spectra of Some "Pure" Minerals

	Na	Mg	Al	Si	S	K	Ca	Fe
Quartz				1.00				
Orthoclase			0.095	0.620		0.285		
Illite	0.077		0.140	0.635		0.148		
Albite	0.080		0.161	0.759				
Kaolinite	0.028		0.472	0.501				
Montmorillonite			0.179	0.727				0.095
Dolomite		0.243					0.736	0.020
Pyrite					0.506			0.494
Gypsum		0.125			0.415		0.460	

small amount of aluminum and perhaps some iron and sulfur associated
with it. However, since XRD verifies the presence of quartz as a
major constituent, we accept the vector. Orthoclase also is a good
fit as shown by Table V. For gypsum, there is considerable more
uncertainty. However, its inclusion added only a small contribution
to the total mass of gypsum. We used these four minerals with the
compositions of the pure minerals listed in Table IV to reproduce
the data for all 107 samples of the albite cluster as per Equation
5. Results for this data reproduction are in Table VI. (The small
discrepancies between observed concentrations in Table VI and Table
II are caused by rounding errors.) On the average, all elements
were well approximated by these four components except sulfur, iron,
and calcium. This suggests that the gypsum component used probably
represents noise. The coefficient of variation for sulfur in this
cluster was 96% and that for iron was 85%. Thus, these elements are
present at very low concentrations and have large standard deviations in this cluster, so we chose to ignore them because they may
constitute noise in these samples. (A future goal of this research
will be to obtain better estimates of variability in elemental
analyses of geological samples by the electron probe EDX method.)

Having obtained a set of mineral components that satisfactorily
reproduces the data, we have defined the C matrix of Equation 5.
Given X and C, Equation 5 is then used to solve for F. For the
albite cluster with 107 samples and 4 mineral components, F is 107 X
4 matrix containing the mass fractions of each mineral component in
each sample. Because these mass fractions sum to 1.0 for each
sample, assuming we are accounting for all the mineral matter, we
solve an overdetermined set of simultaneous equations of the form

$$1.0 = \sum_1 s_1 \, f_{1i} \qquad (6)$$

by least squares regression methods, where 1 is summed over
the mineral components in sample i (albite, quartz, orthoclase, and
gypsum for cluster two). There is a similar equation for each of
the 107 samples. These scaling values are applied to each f_{1i} from
Equation 5 to give the contribution of each mineral component to
each sample. A sum over all samples (i) in each cluster of the form

$$f_1 = \sum_i s_j \, f_{1i} \, w_i \qquad (7)$$

gives the mass fraction f_1 of mineral 1 in a particular cluster. As
Equation 7 indicates, these are corrected by w_i to account for the
variable amount of mineral matter in each sample. This corrects for
the normalization of Equation 1; that is,

$$w_i = \sum_k x_{ik} \qquad (8)$$

where x_{ik} are the original unnormalized EDX intensity ratios. This
assumes that the sum of the measured inorganic components is approximately the same in the absence of organic matter. This is
obviously a minor shortcoming of the method, and better means of
quantifying the inorganic contribution are being implemented.

The f_1 for each cluster are weighted by the number of samples
in the cluster, Equation 9, and the final estimate for the mass

Table V. Best Fit Vectors from Testing Mineral Component
Vectors in Table IV

Element	Vector 1 Albite	Vector 2 Quartz	Vector 3 Orthoclase	Vector 4 Gypsum
Na	7.7e-2			-6.6e-2
Mg				7.6e-2
Al	1.6e-1	6.8e-2	1.1e-1	1.5e-1
Si	7.6e-1	9.8e-1	6.2e-1	2.5e-1
S		4.9e-2		2.5e-2
Cl				
K			2.5e-1	-4.6e-2
Ca		4.1e-2		4.1e-1
Ti				
Fe		8.2e-2		-6.2e-2

Table VI. Average Elemental Contributions of the Mineral Components
Used to Reproduce the Data in the Albite Cluster

	Albite	Quartz	Ortho	Gypsum	Total Predicted	Observed	% Error
Na	0.29e-1				0.29e-1	0.17e-1	6.4
Mg				0.15e-1	0.15e-1	0.11e-1	15.9
Al	0.59e-1		0.12e-1		0.72e-1	0.78e-1	12.3
Si	0.28e+0	0.38e+0	0.81e-1		0.74e+0	0.74e+0	0
S				0.51e-1	0.51e-1	0.23e-1	99.0
Cl						0.12e-2	0
K			0.37e-1		0.37e-1	0.35e-1	4.4
Ca				0.56e-1	0.56e-1	0.83e-1	45.9
Ti						0.19e-2	0
Fe						0.15e-1	0

fraction of each mineral component in the total sample is obtained
by normalizing the sum of the mass fractions to unity, Equation 10.

$$f_1' = \sum_j N_j \, f_1 \qquad (9)$$

where N is the number of samples in cluster j and there is a similar
equation for each mineral component 1. This final normalization is

$$f_1'' = \frac{f_1'}{\sum_1 f_1'} \qquad (10)$$

.

   Our current estimate for the quantitative mineral composition
of the entire oil shale sample based on all 10 clusters is presented
in Table VII. These estimates are consistent with the qualitative
XRD results of Table I. Because they are subject to several sources
of uncertainty, it is impractical to assign error bounds at this
time. These include uncertainty in values of chemical elements for
test vectors, problems in identifying minor mineral components in
the clusters, uncertainty in the relative concentrations of each
element, and uncertainty in the organic content of each sample.
Much of our future research in development of this method will be
aimed at overcoming these uncertainties.

Summary

   The application of multivariate data analysis to the interpretation of chemical element spectra from an electron microprobe-
energy dispersive spectrometer is proving to be a useful method for
quantifying mineral components in complex geological materials. The
method involves the EDX analysis of the solid, cluster analysis of
the chemical element spectra, TTFA followed by determining the
fractional contribution of each mineral component to each analyzed
area by multiple regression, and finally determination of the bulk
composition by summation over all analyzed areas. These techniques
have been applied to an oil shale solid, and the results are
consistent with x-ray diffraction determination of the mineralogy.
Future activities will focus on the generation of a "pure" mineral
spectral library, the determination of uncertainties in the compositional results, the use of multivariate least squares methods to
eliminate the multiple data analysis steps, and method validations.
   Results of these studies are very encouraging and indicate that
a reasonably fast, accurate, and practical method for the quantitative determination of minerals in complex solids can be achieved
with this approach, particularly if multivariate least squares curve
fitting methods can be automated.

Table VII. Quantitative Estimates of Percent Mineral Components in Mineral Matter of Oil Shale Sample[a]

Cluster	Qtz	Ill	Dol	Kaol	Orth	Alb	Gyp	Pyr	Other
1	6.0								
2	4.6		8.7		1.4	3.9	0.8		
3			4.3						
4		2.0		0.3			0.8		0.4[b]
5									1.8[c]
6	7.6		9.0	5.0	4.2				
7			0.8	0.4					
8		1.2		1.5			0.7		0.6[d]
9	4.0	5.6					1.3		
10	0.2	19.0		1.4				0.2	
Total	22.4	27.8	22.8	8.6	5.6	3.9	3.6	0.2	2.6

[a] Dawsonite was not found.
[b] Iron oxide.
[c] $Ca(Mg,Fe)CO_3$.
[d] NaCl, iron oxide.

Literature Cited

1. Lee, R. J.; Huggins, F. E.; Huffman, G. P.;   Scanning Electron Microscopy 1978, 1, 561-568.
2. Huggins, F. E.; Kosmack, D. A.; Huffman, G. P.; Lee, R. J. Scanning Electron Microscopy 1980, 1, 531-540.
3. Howery, D. G. In "Chemometrics: Theory and Application"; Kowalski, B. R., Ed.; American Chemical Society: Washington, D. C., 1977; Chap. 4.
4. Joreskog, K. G.; Klovan, J. E.; Reyment, R. A. "Geological Factor Analysis"; Elsevier Scientific Publishing Company: New York, 1976, Chap. 5.
5. Lawton, W. H.; Sylvestre, E. A. Technometrics 1971, 13, 617-633.
6. Roscoe, B. A.; Chen, C. Y.; Hopke, P. K. Anal. Chimica Acta 1984, 121-134.
7. Dixon, W. J.; Brown, M. B. "BMDP-79 Biomedical Computer Programs P-Series"; University of California Press: Berkeley, 1979; pp. 684.1-684.8.
8. Massort, D. L.; Kaufman, L. "The Interpretation of Analytical Chemical Data by the Use of Cluster Analysis"; John Wiley and Sons: New York, 1983.
9. Malinowski, E. R.; Howery, D. G. "Factor Analysis in Chemistry"; John Wiley and Sons: New York, 1980.
10. Alpert, D. J.; Hopke, P. K. Atmos. Environ. 1981, 15, 675-687.
11. Hopke, P. K.; Alpert, D. J.; Roscoe, B. A. Computers and Chemistry 1983, 7, 149-155.
12. Malinowski, E. R.; Howery, D. G. "Factor Analysis in Chemistry"; John Wiley and Sons: New York, 1980; pp. 72-86.
13. Malinowski, E. R.; Howery, D. G. "Factor Analysis in Chemistry"; John Wiley and Sons: New York, 1980; pp. 88-97, 129-137.

RECEIVED July 17, 1985

Application of Pattern Recognition to High-Resolution Gas Chromatographic Data Obtained from an Environmental Survey

John M. Hosenfeld and Karin M. Bauer

Midwest Research Institute, Kansas City, MO 64110

The application of pattern recognition to a complex chromatographic data base is described. Soil sample extracts were analyzed by high resolution gas chromatography/flame ionization detection (HRGC/FID). The peak retention times were converted to a peak index which was then examined by principal component analysis. Several linear combinations of the peaks were identified as factors which separated the sludge-treated and untreated garden soils. Vector of change plots were constructed that showed the effect of sludge treatment. This data interpretation was achieved without prior knowledge of chromatogram peak identity for either compound class or type.

In a typical environmental survey, a list of target analytes is usually defined in the design phase of the study and prior to sample collection. These analytes may have been chosen through knowledge about the system being studied (1,2), through related environmental situations, or perhaps even by using the analytes currently in vogue, such as priority pollutants (3) or PCBs (4). Each of these approaches, although it may meet the immediate needs of the study at hand, advances the knowledge of the environmental system being studied only to a limited extent. The use of predesignated analytes restricts the information that can be obtained from the samples collected. If indeed the study is designed so that the samples are collected in a statistically determined manner and yet only a small number of target compounds are included for analysis, then the results and probably the study conclusions will reflect this narrow approach.

An alternative approach is to analyze the samples using procedures or instrumentation that will give the maximum amount of data for each sample. For example, recent advances in atomic spectroscopy, i.e., inductively coupled argon plasma emission spectroscopy (ICP-AES), allow 20 to 30 elements to be detected simultaneously.

0097–6156/85/0292–0069$06.00/0
© 1985 American Chemical Society

Another means of greatly increasing the amount of data on the organic
compounds present in samples is through the use of universal techniques such as high resolution gas chromatography (5,6) combined with
flame ionization detection (HRGC/FID).

The problem is to sort through and retrieve information from
the large amount of quantitative data produced with capillary chromatography. After those peaks that contain the most information that
describes the sample have been determined, directed and specific confirmation analysis by GC/MS may occur.

In order to illustrate this concept, the use of pattern recognition approach on gas chromatographic data will be presented. This
paper will focus on an environmental survey of sewage sludge usage
on home vegetable gardens. The analysis of the organic content of
the soils collected on this survey was an opportunistic study since
the original purpose was to monitor trace metal levels in the treated
and untreated gardens. The addition of the HRGC/FID analysis will
hopefully add to the knowledge base.

Experimental

Soil samples were collected from 92 gardens as part of a nationwide
survey of the usage of sewage sludge on home vegetable gardens. In
each designated county, soil was collected from each of two garden
types, i.e., sludge treated and untreated. Samples of sludge, when
available, were also collected from the garden sites.

In the laboratory, each soil sample (40 g) was transferred to a
centrifuge bottle. Since the original purpose of the soil collection was to monitor specific organic compounds in the sludge-amended
garden soils, a set of surrogate compounds was added to the soil
prior to extraction to assess the extraction and cleanup recovery.
The surrogate compounds were mono-, tetra-, octa-, deca-^{13}C-PCBs,
d_8-naphthalene, ^{13}C-PCP and ^{13}C-phenol. The soil samples were dried
with Na_2SO_4 (60 g) and then Soxhlet extracted with hexane:acetone
(9:1) for 16 h. The extract was dried with sodium sulfate, concentrated, and split. While one portion was held for other analyses,
the other portion was placed on a 3% deactivated silica gel column
and eluted with increasing solvent polarity systems [hexane, followed by methylene chloride:hexane (1:1), and then methylene
chloride:acetone (95:5)]. The extracts were combined and reduced
to 1 mL, split and two internal standards added (tetrafluorobiphenyl
and d_{12}-chrysene). The extracts were chromatographed on a 15-m DB-5
fused silica capillary column and detected with flame ionization
(FID). Sludge samples were extracted according to the EPA sludge
protocol (7) developed at Midwest Research Institute.

The output from the FID was captured by a Nelson Chromatographic
Data System and stored on floppy disks. The algorithm in the data
system processed the raw chromatograms and stored the peak retention
times and areas in data tables which were subsequently transferred
to a Digital Equipment Corporation (DEC) 11/23+ for further processing. A relative retention index (8,9) was developed on the internal
standards added to each sample. An arbitrary chromatogram was chosen
to act as a reference against which the other chromatograms would be

compared and the peak numbering system developed. This procedure
consisted of lining up the two internal standards respectively across
all chromatograms. Within a given chromatogram, each retention time
(X) was transformed to obtain a retention time (Y) using the simple
linear equation:

$$Y = aX + b \qquad (1)$$

such that

and

$$aS_{i1} + b = S_{r1} \qquad (2)$$

$$aS_{i2} + b = S_{r2} \qquad (3)$$

where S_{r1} and S_{r2} = retention times of the two internal
                              standards in the reference chromatogram

and S_{i1} and S_{i2} = untransformed retention times of the
                              two internal standards in chromatogram i.

The system of equations 2 and 3 yields:

$$\text{the slope } (a) = \frac{S_{r2} - S_{r1}}{S_{i2} - S_{i1}} \qquad (4)$$

and

$$\text{the intercept } (b) = \frac{S_{r1}S_{i2} - S_{r2}S_{i1}}{S_{i2} - S_{i1}} \qquad (5)$$

Thus, within each chromatogram, each retention time (X) was linearly
transformed using Equations 1, 4 and 5 to obtain the adjusted retention time (Y). Next, the retention time of the first internal standard was renumbered as peak index 1. Peaks occurring prior to the
first internal standard were deleted in each chromatogram because
they were on the solvent peak. Using a 4-sec peak retention window,
each retention time in subsequently adjusted chromatograms was numbered based on the window in which it occurred. When two peaks in a
given chromatogram were less than 4 sec apart and within the same
window, the peaks were assumed to be unresolved and therefore summed
(this happened 14 times out of a total of over 10,000 peaks in the
entire data set).
    Pattern recognition, i.e., principal components analysis, was
attempted on the data matrix of 92 chromatograms x 364 peaks. However, the mathematical requirements of the Statistical Analysis System (SAS) specify that the number of observations (chromatograms) be
greater than the number of features (peaks) for matrix inversion
computations. To solve this problem we considered (1) dividing the
chromatogram into three or four sections containing an equal number
of peaks or (2) considering sets of 91 randomly selected peaks in an
iterative process. However, a significant drawback of these two approaches is that any interrelationships which may exist between different portions of the chromatogram are not taken into account.

An alternate approach (10) was to artificially increase the 92 x 364 matrix to a 368 (4 x 92 chromatograms) x 364 (peaks) so that the matrix could be inverted and all 364 peaks considered simultaneously. This was done by replicating the original 92 observations and by slightly modifying the peak areas in each replicate (the area values were multiplied by a random number between 0.99 and 1.01). The entire chromatogram was analyzed rather than portions of it and thus the correlations between peaks were preserved. In the first pattern recognition step, a principal component analysis using SAS was performed on the combined data set of 368 chromatograms with 364 peaks each. This procedure yielded 364 factors, with the first three explaining 16.2%, 10.5%, and 6.9% of the total variance, respectively. Within these three factors, those peaks with high loadings were kept until a maximum set of 91 peaks was retained, such that factor 1 contributed 66 peaks, factor 2, 31 peaks and factor 3, 6 peaks. Then using these 91 peaks only, the original data set was reexamined by principal components analysis. Eigenvalues greater than one were plotted to determine how many factors should be retained. After varimax rotation, the factor scores were plotted and interpreted.

Results and Discussion

Typical chromatograms of soil extracts are shown in Figure 1. It can be seen that the chromatograms are complex and that the sludge-treated soil sample has a greater number of peaks ($\sim$ 150 vs. $\sim$ 50) and higher detector response than the untreated soil sample. One might anticipate that there is a structure in the data set of treated and untreated soils and that this structure might be resolved by application of pattern recognition techniques.

    The analysis of chromatographic data is usually performed on normalized chromatograms, which is an attempt to account for the mass injected. However, the closure of analytical data is a problem with normalized data which has been described elsewhere (11). We examined our data for this problem by plotting the grand mean variation over all 368 peaks versus the standard deviations of these peaks. Closure did not occur in the unnormalized data.

    The plot of the decreasing sequence of eigenvalues of the 91 principal components is shown in Figure 2. Components 1, 2 and 3, with eigenvalues of 42.7, 22.4 and 8.8, respectively, explained 47.0%, 24.6% and 9.7% of the total variance, respectively, a total of 81.3%. The fourth component with an eigenvalue of 2.5 accounted for only 2.7% of the total variance, and thus only the first three principal components were selected to be further explored. (Note that only 9 of the 91 components had eigenvalues greater than 1.0, explaining together 92.3% of the total variance.) After varimax rotation, the eigenvalues of the first three components were only slightly changed to 42.1, 20.9 and 10.8, respectively; thus a strong first factor remains followed by two factors approximately half as important as their precedent. Next, within each factor, those features (peak numbers) with loadings representing at least 2% (about twice the average of 1/91 $\cdot$ 100%) of the variance of this factor were kept and ordered with respect to these percentages. By this method,

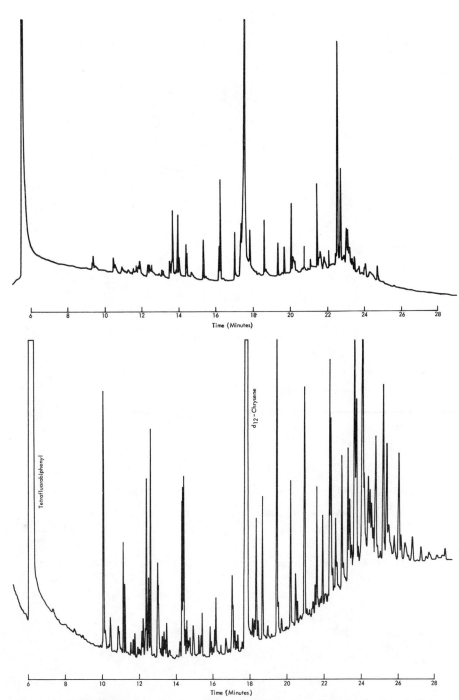

Figure 1. Typical gas chromatograms of soil from an untreated garden (top) and sludge treated garden (bottom). Conditions: 15 m DB–5, 0.25 mm ID capillary column operated at 100 C (2 min) then programmed at 10 C/min to 310 C (7 min hold).

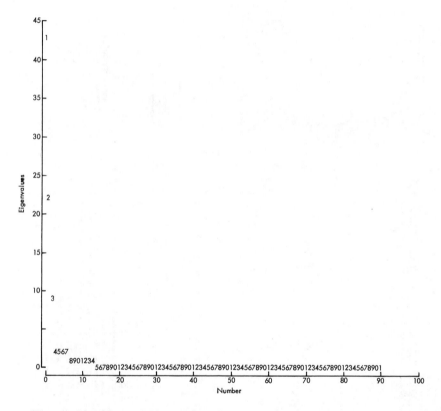

Figure 2. Plot of the eigenvalues of the correlation matrix.

Factor 1 was characterized by 25 peaks, each having about equal loadings (0.99 to 0.93 or equivalently 2.3% to 2.0% of the variance accounted by this factor). The first plot in Figure 3 shows that these
25 peaks are spread across the total peak range with a somewhat
higher concentration of peaks toward the end of the range (276 to
351). In contrast, Factor 2 contained 24 peaks with loadings ranging from 0.97 to 0.71, representing proportions of variance of this
factor from 4.5% to 2.4%. These 24 peaks were somewhat clustered
between peak numbers 31 and 80, as shown in the second plot in Figure
3. Factor 3 (3rd plot in Figure 3) is the most striking in comparison to the previous two factors. A small number (14 out of 91) of
peaks account for 84.8% of the variance explained by this factor,
with loadings ranging from 1.0 to 0.6 (equivalent to high percent
variances ranging from 8.8 to 3.3%). Although these 14 peaks broadly
cover the whole range of the original 400 peaks, a minor cluster occurs in the 192 to 248 section. It is interesting to note that only
one peak, number 161, is duplicated in any of the factors (2 and 3),
thus underlining the orthogonality of these three factors to each
other. These loading variance percentage plots indicate that a
structure may be present in the data set due to the above discussed
dissimilarities.

     In order to determine the scope of the hidden structure, factor
plots of the observations were made. A plot was made of the factor
scores for each garden soil, i.e., sludge-treated (T) or untreated
(U). No clear pattern emerged from these plots of the factor scores
and so another approach was taken. The treated and untreated scores
were replotted (Figure 4) with a letter code substituted for each
county from which a soil sample was collected. However, these plots
were only of minor use in providing insight into the data structure.
From these factor plots of the observations, secondary plots were
constructed to determine the effect of treating garden soil with
sludge compared to untreated garden soil. These vectors of change
plots are shown in Figures 5, 6, and 7. It is important to emphasize
again that the treated and untreated soil sample came from at least
two separate gardens within a county, i.e., no experimental design
of adding sludge to untreated gardens occurred. Figure 5 presents
the plot of factor 1 versus factor 2. The comparison of sludge to
untreated soils M-Z, G-T, B-O, and A-N shows an equal and positive
combination of factors 1 and 2. Site G-T is profoundly affected by
sludge treatment, as evidenced by the large response in these two
factors. Soils J-W, L-Y, K-X, E-R are negatively affected by factor
2. In Figure 6, a similar effect is seen for soils G-T, A-N, and
M-Z; however, site C-P is reversed from Figure 5 because of a strong
contribution from factor 3. Sites D-Q, I-V, F-S have the same rever-
sal as site C-P. Sites E-R, K-X, and L-Y are strongly affected by a
positive factor 3. Figure 7 shows changes similar to those occurring
in Figure 6. It is apparent that the relationship among factors is
3 > 2 > 1. However, it is important to recognize that these eigenvector projections were made without knowledge about the class assignments of the individual soils. The resulting separation is
therefore a strong indication of real differences between the two
garden soil types in a given county. Similar vectors of change were

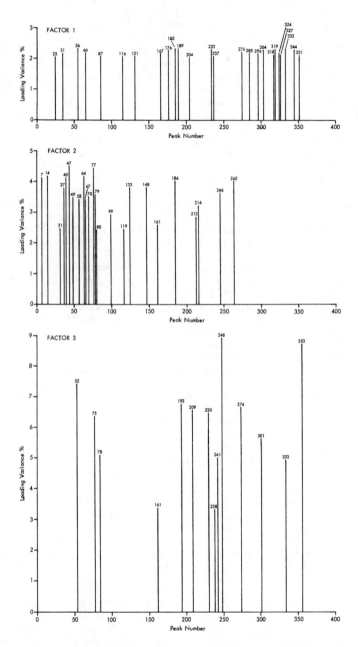

Figure 3. Plot of the features (peak numbers) compared to the loading variance percent for the first three factors.

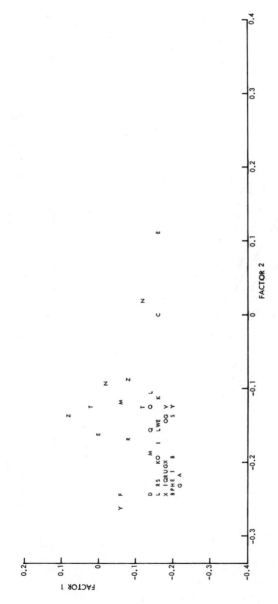

Figure 4. Factor score plots of sludge treated and untreated garden soil.

(Letters A-M are untreated soil while N-Z are the corresponding sludge treated soil such that A-N are both types of soil from a given county.)

15 scores not plotted because of superposition.

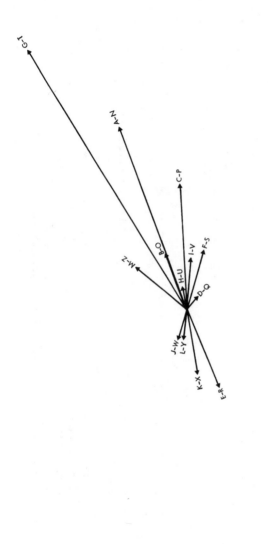

Figure 5. Vector of change for factors 1 versus 2 in untreated and sludge treated garden soil by county sampled.

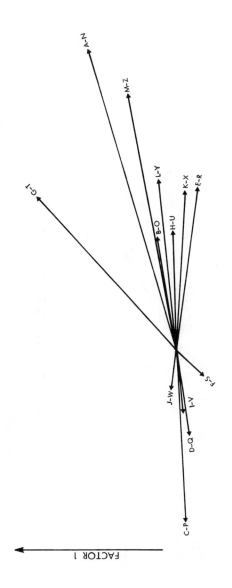

Figure 6.   Vector of change for factors 1 versus 3.

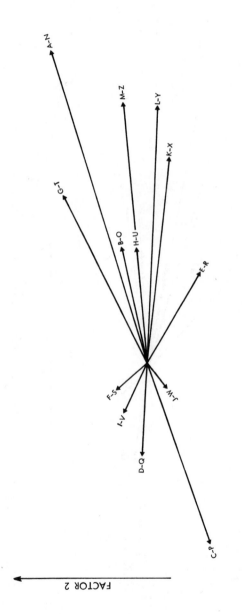

Figure 7. Vector of change for factors 2 versus 3.

seen in the principal components analysis of trace metal data from these same soils (12). These plots also showed the effect of sludge treatment on trace metal levels of garden soils as indicated by the vectors of change.

Conclusions

A means of abstracting the most relevant information from a chromatographic data set has been presented. Principal components analysis has been shown to be a powerful technique for obtaining the structure hidden in a complex data set. The merits of this procedure are its usefulness in pointing out (1) the chromatographic peaks that require further compound identification and (2) the peaks that exhibit similarity or dissimilarity, which lead to data set insight enhancement. Specific compound identification time using gas chromatograph/mass spectrometry may be minimized because the peaks causing the effect in the soils have been determined. This leads to less GC/MS identification time and thus possible lower cost of analysis. The correctness of the approach, however, applied to the present data set will be verified when the compound identity of the peaks is known and this identity leads to an understanding of the effects of sludge treatment on garden soils.

Acknowledgments

The authors wish to thank L. Moody for performing the data transfer from the Nelson chromatographic data system to the DEC 11/23+ and D. Harwood for performing the closure analysis.
     The soil samples were obtained and analyzed on EPA Contract No. 68-01-5915, Task 42, D. T. Heggem, Task Manager. Pattern recognition work was funded by a Midwest Research Institute internal fund, RA-321.

Literature Cited

1. Popham, J. D.; D'Aurla, J. M. Environ. Sci. Technol., 1983, 17, 576-82.
2. Kvalheim, O. M.; Øygard, K.; Grahl-Nielsen, O. Anal. Chim. Acta, 1983, 150, 145-52.
3. Giabbai, M.; Roland, L.; Chian, E. S. K. In "Chromatography in Biochemistry, Medicine and Environmental Research"; Frigerio, A., Ed.; Elsevier: Amsterdam, 1983; Vol. 1, p.41.
4. Dunn, W. J., III; Stalling, D. L.; Schwartz, T. R.; Hogan, J. W.; Petty, J. D.; Johansson, E.; Wold, S. Anal. Chem., 1984, 56, 1308-13.
5. Hsu, F. S.; Good, B. W.; Parrish, M. E.; Crews, T. D. J. HRC & CC, 1982, 648.
6. Clark, H. A.; Jurs, P. C. Anal. Chem., 1975, 47, 374.
7. Haile, C. L.; Lopez-Avila, V. "Development of Analytical Test Procedures for the Measurement of Organic Priority Pollutants in Sludge," EPA-600/S4-84-001, USEPA, March 1984.
8. Mayfield, H. T.; Bertsch, W. Computers in the Analytical Laboratory.

9. Koskinen, L. Trends in Analytical Chemistry, 1982, 1, 324.
10. Meglen, R., personal communication.
11. Johansson, E.; Wold, S.; Sjödin, K. Anal. Chem., 1984, 56, 1685-8.
12. Hosenfeld, J. M., unpublished data.

RECEIVED September 18, 1985

6

Quality Assurance Applications of Pattern Recognition to Human Monitoring Data

Philip E. Robinson, Joseph J. Breen, and Janet C. Remmers

Office of Toxic Substances, U.S. Environmental Protection Agency, Washington, D.C. 20460

Principal Component Analysis (PCA) is performed on a human monitoring data base to assess its ability to identify relationships between variables and to assess the overall quality of the data. The analysis uncovers two unusual events that led to further investigation of the data. One, unusually high levels of chlordane related compounds were observed at one specific collection site. Two, a programming error is uncovered. Both events had gone unnoticed after conventional univariate statistical techniques were applied. These results illustrate the usefulness of PCA in the reduction of multi-dimensioned data bases to allow for the visual inspection of data in a two dimensional plot.

Data have been collected since 1970 on the prevalence and levels of various chemicals in human adipose (fat) tissue. These data are stored on a mainframe computer and have undergone 'routine' quality assurance/quality control checks using univariate statistical methods. Upon completion of the development of a new analysis file, multivariate statistical techniques are applied to the data. The purpose of this analysis is to determine the utility of pattern recognition techniques in assessing the quality of the data and its ability to assist in their interpretation.

Background

Under the Toxics Substances Control Act, the Environmental Protection Agency (EPA) is mandated to gather data on the exposure of the general population to toxic substances. Toward this end, the Office of Toxic Substances within the EPA has undertaken several long term monitoring programs. These programs involve the collection of human tissue specimens from a statistically representative

sample of the United States population and the subsequent chemical
analysis for a select group of toxic substance residues and
their metabolites. The data generated by these studies are
used to establish the prevalence and levels of human exposure,
to identify trends in this exposure, and to assess the effects
of regulatory action on exposures to these chemicals.

The National Human Adipose Tissue Survey (NHATS) (1) is
an on-going program conducted annually since 1970. Human adipose
tissue specimens are collected during either post-mortem examinations
or elective surgical procedures at 40 locations across the continental United States. Demographic characteristics of each tissue
donor are also reported. Since the program's inception, over
20,000 specimens have been chemically analyzed at seven analytical
laboratories. The adipose tissue specimens are chemically analyzed
using a packed column gas chromotography/ electron capture detector
method and the Mills Onley Gaither procedure (2). Data were
gathered on 19 organochlorine compounds and PCB's. A list of
the residues measured in adipose tissue is found in Table I.

TABLE I. Chemical Residues Measured in Adipose Tissue

p,p' DDT	Aldrin
o,p' DDT	Dieldrin
p,p' DDE	Endrin
o,p' DDE	Heptachlor
p,p' DDD	Heptachlor Epoxide
o,p' DDD	PCB's
alpha BHC	Oxychlordane
beta BHC	Mirex
gamma BHC	trans-Nonachlor
delta BHC	Hexachlorobenzene

The survey design used by NHATS is based on a multi-stage
selection process in which the first stage involves the random
selection of a specified number of population centers (SMSA's)
from each geographical region of the country. At the second
stage, a local medical examiner or pathologist from each SMSA
is identified and asked to contribute tissue specimens according
to demographic quotas based on age, race and sex.

This study provides EPA with human monitoring data to assess
the level of exposure of the general population to various toxic
substances. Statistical analyses of these data have primarily
involved a description of the distribution of these chemicals
in the population. Specifically, the proportion of specimens
for which a particular residue level was quantified and the
level of the chemical detected have been reported for various
age, race, sex and geographical strata.

Approach

Exploratory data analysis (3) is performed on the data base
using multivariate statistical techniques. The objectives of

this analysis are to assess the applicability of pattern recognition
techniques in the quality control of human monitoring data and
to assess its ability to identify relationships between variables
contained in the data base.

Previous analyses were confined to the use of univariate
techniques applied to the individual chemical residue levels.
In contrast, this analysis focuses on the evaluation of relationships
between and among all quantitative variables simultaneously.
To simplify the effort, various subsets of the data base are
examined. The intent of this action is to allow for model validation
or confirmation should relationships of interest be identified.
The initial data set consisted of 3800 records relating to specimens
collected during the years 1977 to 1981 for those chemical residues
having a greater than 10% detection rate. Table II lists those
variables and residues included in the analysis. As the analysis
progressed changes were made to this data set to either facilitate
interpretation of the results or to further investigate hypotheses
generated by the data.

TABLE II. Variable List of Initial Data Set

Variable Name	Residue
Date of Collection	Hexachlorobenzene
Date of Analysis	trans-Nonachlor
Lab Code	Oxychlordane
Geographical Region	p,p' – DDT
Age	p,p' – DDE
Sex	alpha Benzene Hexachloride
Race	beta Benzene Hexachloride
Length of Storage	gamma Benzene Hexachloride
Medical Diagnosis Code	Heptachlor Epoxide
	Dieldrin
	PCB's

An examination of summary statistics was conducted to determine
which variables to include in the initial analysis. Measures
of association between variables, i.e., correlations, were investigated to ensure that a high degree of multicollinearity did
not exist between any pair of variables. The need for data
scaling, transformation, or dimensionality reduction was also
evaluated. For example, body burden data tend to be lognormally
distributed. Whether these data need to be transformed prior
to using techniques such as principal component analysis (PCA)
is critical to the development of a basic strategy for the analysis
of this and other data sets containing human monitoring data.

The initial multivariate analysis consisted of a principal
component analysis on the raw data to determine if any obvious
relationships were overlooked by univariate statistical analysis.
The data base was reviewed and records containing missing data
elements were deleted. The data was run through the Statistical
Analysis System (SAS) procedure PRINCOMP and the results were
evaluated.

Results

Figure 1 is a plot of the first two principal components (PC1 and PC2) of data from collection year 1981. The symbols on the plot (1 to 9) designate the Census Division number in which the specimen was collected (1=NE, 2=MA, 3=ENC, 4=WNC, 5=SA, 6=ESC, 7=WSC, 8=Mo, 9=Pa). A map of the continental United States which graphically illustrates the Census Divisions is provided in Figure 2. A striking observation from Figure 1 is the concentration of 6's on the left of the plot. A 6 represents the South Central Census Division. Table III lists the loading factors associated with the first and second principal components for several chemical residues included in the analysis. The negative loadings provided by the chlordane-related compounds, Oxychlordane and Heptachlor Epoxide, in the second principal component are of specific interest since they are in direct contrast to the other loading scores. Subsequent analysis of these data found that high levels of these compounds were confined to one sampling location within the Census Division. Two possible explanations for this phenomena are (1) the samples were contaminated at the collection site or (2) there exists an exposure problem to these chemicals in this geographical area. The Census Division 6 data were produced over an extended period of time and include results with non-elevated levels of chlordane-related compounds. This suggests that the specimens and not the laboratory are the source of the problem. An investigation is being conducted by the EPA to determine the cause of these levels.

TABLE III. Residue Principal Component Loading Factors
Data is from Collection Year 1981

Variable	PrinComp 1	PrinComp 2
p,p' DDT	.414	.075
p,p' DDE	.437	.169
alpha BHC	.009	.033
beta BHC	.395	.006
gamma BHC	.039	.067
Dieldrin	.144	.155
Oxychlordane	.465	-.104
Heptachlor Epoxide	.268	-.335

Figure 3 is another plot of output from the PRINCOMP procedure on data collected between 1977 and 1979. The symbols on the plot represent the year of collection of the specimens (7=1977, 8=1978, 9=1979). A pattern related to the dispersion of the 7's, 8's, and 9's is visible but any conclusion at this point is tentative due to the large number of hidden (unplotted) observations. Examination of the loadings for principal components 1 and 3 (PC1 and PC3) in Table IV note the contribution of the residues, p,p' DDE and p,p' DDT, to principal component 3. To better assess the effect of these variables on the group

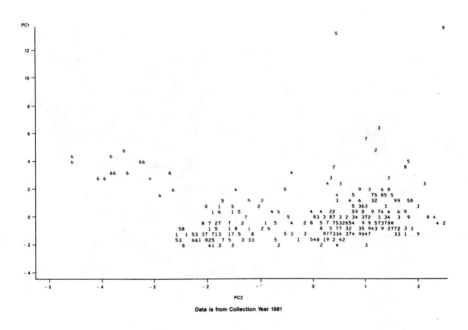

Figure 1.   Plot of PC1 vs. PC2.   (Symbol is Number of Census
Division)

Figure 2. U.S. Census Divisions

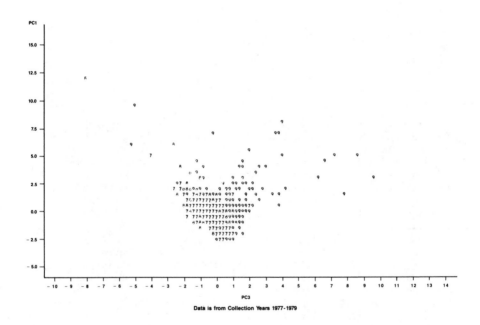

Data is from Collection Years 1977-1979

Figure 3. Plot of PC1 vs. PC3 - Uncorrected Data (Symbol is
Year of Collection: 7=1977, 8=1978, 9=1979)

of 9's outside the central cluster, the scale on the plot was
changed and the data replotted. Figure 4 is the revised plot.

TABLE IV. Residue Principal Component Loading Factors
           Data is from Collection Years 1977 - 1979

Variable	PrinComp 1	PrinComp 3
p,p' DDT	.358	.452
o,p' DDT	-.012	.064
p,p' DDE	.410	.419
o,p' DDE	.021	-.036
beta BHC	.078	.272
Dieldrin	.124	-.083
trans-Nonachlor	.376	-.240
Oxychlordane	.453	-.282
Heptachlor Epoxide	.157	-.171
Hexachlorobenzene	.082	.305
PCB's	.294	-.351

The concentration of 9's in the right side of the plot
in Figure 4 indicates a potential bias in the 1979 data for
those variables with the large positive loading scores in principal
component 3. In an effort to explain these factors, the data
were sorted by the value of the third principal component and
a printout of the data was examined. The majority of the high
scores for PC3 were associated with specimens collected in 1979.
Further analysis indicated that the p,p' DDE residue levels
are unusually high for a large number of specimens in this year.
Although, individually, each of the data points passed range
checks normally used to screen for outliers, the frequency of
such high levels is highly unlikely given the wide variety of
demographic and geographic strata from which these specimens
were collected. In addition, as these specimens were chemically
analyzed over the course of a year, the problem could not have
resulted from the analytical technique used to quantify these
levels.
    A review of the raw data resulted in the discovery of an
error in the computer program which created the analysis file.
All residue levels greater than 1.0 were coded in the analysis
file with an extra 0 between the decimal point and the first
unit's place. For example 2.46 was recorded as 20.46. The
limited number of such levels did not significantly affect previously
computed univariate statistics and these artificial outliers
remained undetected. Figure 5. presents a plot of the PRINCOMP
output after the analysis file was corrected. This plot shows
a more uniform distribution of data points for specimens collected
in each of the three years.

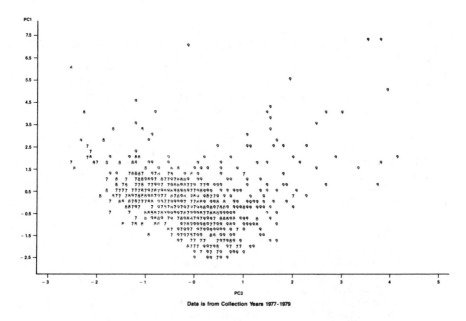

Figure 4. Rescaled Plot of PC1 vs. PC3 - Uncorrected Data
(Symbol is Year of Collection: 7=1977, 8=1978, 9=1979)

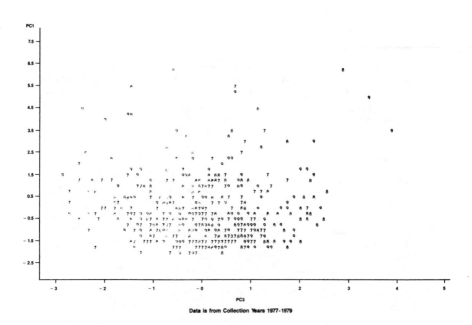

Figure 5. Plot of PC1 vs. PC3 - Corrected Data (Symbol is
Year of Collection: 7=1977, 8=1978, 9=1979)

Conclusions

This preliminary analysis of human monitoring data has identified
two significant situations that had gone undetected after conventional data checks were made. It should not be concluded, however,
that other techniques could not provide this identification.
The relative ease at which non-statisticians can make use of
a sophisticated technique such as Principal Component Analysis
speaks to its power in the hands of more accomplished practitioners
or chemometricians. Simple univariate analyses are not sufficient
to adequately check the large volume of data coming from state-of-
the-art chemical analytical procedures.

For example, a single estimate for total PCB's has been
historically collected in the NHATS program. Current advances
in chemical analysis protocols now allow for the determination
of isomer specific resolution of PCB's. Given the 209 PCB's
that are now possible to detect, an adequate evaluation of the
data without the use of pattern recognition techniques seems
impossible. From a QA/QC perspective, these methods can facilitate
the detection of outliers and aid in the interpretation of human
chemical residue data. The application of statistical analysis
must keep abreast with these advances made in chemisty. To
handle the complexity and quantity of such data, the use of
more sophisticated statistical analyses is needed.

Work is continuing on the application of pattern recognition
to the human monitoring data base to assist in the identification
and interpretation of potential underlying structures associated
with this data base.

Disclaimer

Literature Cited

1. Mack, G., et. al., Survey Design for the National Human
Adipose Tissue Survey, Battelle Columbus Laboratories, Columbus,
Ohio. Report to the U.S. Environmental Protection Agency, Washington,
D.C., Contract No. 68-01-6721, 1884.
2. Erickson, M.D., Evaluation of Analytical Methods and Data
for Hexachlorobenzene in Human Adipose Tissue, Midwest Research
Institute, Kansas City, Missouri. Draft Report. U.S. Environmental
Protection Agency, Washington, D.C., Contract No. 68-02-3938,
1983.
3. Parsons, M. and Wolff, D., Pattern Recognition Approach
to Data Interpretation, Plenum Press, New York, 1983.

RECEIVED September 6, 1985

Description of Air Pollution by Means of Pattern Recognition Employing the ARTHUR Program

F. W. Pijpers

Department of Analytical Chemistry, University of Nijmegen, The Netherlands

Pattern recognition methods have been used for the description of air pollution in the industrialized region
at the estuary of the river Rhine near Rotterdam. A
selection of about eight chemical and physical-meteoro-
logical features offers a possibility for a description
that accounts for about 70% of the information that is
comprised in these features with two parameters only.
Prediction of noxious air situations sometimes succeeds
for a period of at most four hours in advance. Sometimes, however, no prediction can be made. Investigations pertaining to the correlation between air composition and complaints on bad smell by inhabitants of
the area show that, apart from physical and chemical
descriptors, other features are also involved that
depend on human perception and behaviour.

Description of the Problem

In our laboratory, pattern recognition has been used in solving a variety of problems. Recently we set for ourselves a goal
to investigate the probability of describing air pollution in highly industrialized regions in such a way that, by taking appropriate
measures, complaints from inhabitants of the area would be prevent-
ed. The DCMR* - a governmental organization in The Netherlands -
collects data on various constituents of polluted air at a number
of locations situated near and in a highly industrialized area at
the estuary of the river Rhine, near Rotterdam. Occasionally, when
weather conditions demand it, warnings for expectations of emergency air pollution situations are dispatched to the industries in the
area. These dictate a limitation of the emission of SO2, resulting
from burning of sulfur-containing fuel. In spite of these well organized actions, complaints from inhabitants of the area, who are
stimulated to communicate their observations by telephone to the
same office that dispatches the warnings, cannot be precluded.

* Central Environmental Control Agency, Rijnmond

0097-6156/85/0292-0093$06.00/0
© 1985 American Chemical Society

Because of the complex situation, which has to be described by a set of parameters pertaining to the constitution of the atmosphere at various locations and to the weather conditions, the application of pattern recognition methods seems obvious. (1-5)

The aim of this investigation is twofold:
- Finding the relevant features that describe emergency situations.
- Prediction of the evolution of these features in time, in order to enable a forecast of potential emergency situations and allow the proper measures to be taken in time. Thus, the burden on the inhabitants of the area may be alleviated.

Without going into the details of the numerous techniques that are being used in pattern recognition, a general outline of the method of problem handling by means of the ARTHUR package may be clearly illustrated from an approach to the air pollution problem. (See Table I)
- For a start, the pattern of an atmospheric composition and situation, i.e., a data vector comprising all available physical and/or chemical data pertaining to that situation, is positioned in a multidimensional feature space that is spanned by all physical (i.e., meteorological) and chemical (i.e., compositional) named features.
- When a number of situations, positioned in that feature space, group together or cluster, it is obvious that their physical and chemical behaviour is similar. This will be perceived by the population of the area in the same way. In pattern recognition it is assumed that such behaviour not only holds for the known physical and chemical data but also reflects similar behaviour of properties such as fresh air or noxious air.
- In this discussion we select a number of consecutive days where a period with many complaints is preceeded and followed by an about equal period of "good" days to see whether at least two different clusters of patterns in the feature space may be found that correspond with the property polluted air versus fresh air.
- In case we succeed in finding these clusters we may proceed by selecting those features that are most relevant for the definition of the clustering. Here the techniques applied focus on correlation among the features themselves and a correlation between the features and the digitized property. (CORREL and WEIGHT are the appropriate sub-routines in the ARTHUR package). (6)
- Finally the relevant feature combination for the description of the situations where complaints may occur can be used to predict the possible occurrence of bad situations in the (near) future.

Discussion of the results

In Figure 1, a map is presented of the estuary of the river Rhine near (west of) Rotterdam. The industrialized region is situated around the harbors located south of the river near Hoogvliet. Refineries and fertilizer plants are found there. In Table 1, various stages in pattern recognition are listed. The subroutine CHANGE, in combination with the INPUT-format is used in stages one and two. HIER, TREE and PLANE are useful in stage three, whereas CORREL and

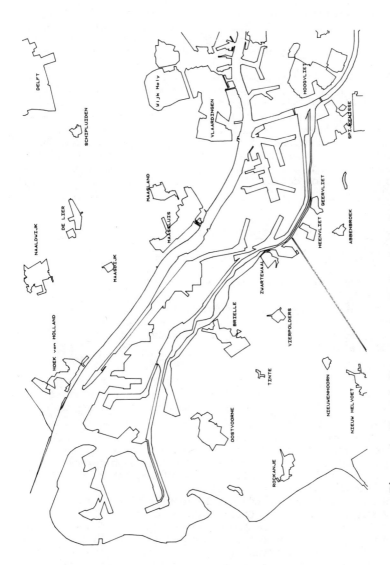

Figure 1. Map of the estuary of the river Rhine, near Rotterdam. Reproduced with permission from Ref. (7). Copyright 1984, The Royal Society of Chemistry.

WEIGHT are employed in stage four. Addition of an extra number of patterns in a TEST-set allows validation of the learning machine developed in the previous stages.

Table II lists the chemical and physical measurements that produce the feature space. A list of complaints as coded from the communications from the population is also given. Application of the inter-feature correlation calculation, CORREL, on the chemical and physical features listed here, results in a limitation of the number of features without sacrificing too much information. Apart from the stability parameter, representing the meteorological conditions, some chemical constituents of polluted air are found to be of importance in describing the situation (see Table III).

Construction of the interfeature variance-covariance matrix, followed by an eigenvector-eigenvalue calculation according to the Karhunen - Loéve procedure (KARLOV), produces a number of eigenvectors equal to the number of features. It is found that the two highest eigenvalues comprise 78% of the total information, and thus, should provide a reasonable picture of the situation. The two clusters representing "bad" and "good" situations resulting from a projection on the plane defined by the first two eigenvectors is given in Figure 2.

Table I. Various Stages in Pattern Recognition

1. Definition of problem space and data vectors

2. Selection of patterns for a training set

3. Search for clusters by various techniques

4. Selection, ordering and weighting of relevant features
   (iteration between step 3 and 4)

5. Predictions for a test set

From this figure one could get the false impression that the problem has been solved. This is not true because this result could only be obtained with a carefully selected data set measured early in May, 1979. The weather was stable during the entire observation period, comprising two weeks with many complaints, followed and preceeded by two weeks of practically no complaints. The other months of that year showed a much less predictable situation.

In order to see whether the development in time of a given situation could be followed, autocorrelation functions of all relevant features were constructed. From these functions it was observed that, provided the weather was not too unstable, an autocorrelation time of about four hours was encountered. This autocorrelation was best defined for SO2 concentrations that are measured hourly at various

Table II. List of Observed Features
_____

Chemical Compounds   Meteorological Conditions

Nitrous oxide   Direction of the wind
Nitric oxide   Speed of the wind
Sulfur dioxide   Relative humidity
Standardized smoke   Temperature
Saturated and   Sunlight radiation
 Olefinic Hydrocarbons   Amount of precipitation
Ozone   Air pressure
                                      Cloudiness
                                      Stability

            Types of Complaint

            Soot, dust
            Noxious smell
            Acid, chemical smell
            Oily smell
            Smog
            Others
_____

Table III. List of Relevant Features
_____

Stability (Meteorological conditions)
Ozone
Saturated hydrocarbons
Olefinic hydrocarbons
Sulfur dioxide at three different locations
_____

locations in contrast to other air constituents that are measured
less frequently. This time dependency has been translated into
features that could be employed in the pattern recognition process
by introducing not only the actual concentration of SO2 but also its
time dependence as concluded from observations up to four hours in
the past.

Based upon these features another learning machine was constructed for the year 1982, describing the situation with exclusion
of the months September till December when the weather was in general
too unstable. From that period 36 situations have been selected where
complaints were registered for at least two consecutive hours. These
data were completed by another 36 situations without complaints, selected exactly 24 hours before or after a period with complaints.

The time dependence of the direction of the wind was taken into account by integration over a period of four hours in the past.
These features were autoscaled, weighted and combined linearly
according to the Karhuhnen-Loeve transformation. (See Table IV).
This table refers to the situation in the city of Schiedam. The
eigenvectors listed here are represented by the squares of the
coefficients; the two most important ones account for 68% of the
variance of the entire set of features. It is seen in Figure 3 that
the projection of the patterns on the plane spanned by these two
eigenvectors is dominated by two complaint patterns with exeptional
feature values. One may also note the overlap between the cluster
with complaints and that without complaints.

In search for a better description, and taking into account
that the impressions registered by the human nervous system should
be rated on a logirithmic scale; a new transformation was tried,
this time based upon logarithmized features. This resulted in Table
V, where it is seen that a somewhat enhanced information from the
first two eigenvectors was obtained, in comparison with that of
Table IV (68% and 71% respectively). The merits of this learning
machine are visualized in Figure 4, where apart from the training
set patterns an extra set of 13 patterns with complaints and 13
patterns without complaints is added and projected as a test set.
According to the nearest neighbour voting system, eight out of
thirteen non-complaint situations and nine out of thirteen complaint
situations are positioned correctly. This is not an impressive
result. Apparently the "reason" for complaints is not exclusively
described by physical or chemical parameters.

This observation is also illustrated by another calculation.
In Figure 5, the hourly patterns of one day (24 hours) were projected on the training set. This day, May 17th, 1982, at Schiedam com-
prises two hours with complaints, viz., 13 and 14 hour. It is seen
that the situation evolves from the area where no complaints are
predicted towards the complaint area. About 8.00 hour the borderline
is crossed and indeed at 13.00 and 14.00 hour complaints are recorded.
The 15th and 16th hour, that are clearly in the complaint area, are
not signalled, and the trace ends at 23 hours without complaints in
the non-complaint area as was expected.

However, this procedure failed completely with the hourly
data set collected on July 7th and 8th in the same location (See
Figure 6). Here the evolution in time of the air composition pattern
vector circles around in the boundary area between complaint and
non-complaint situations. There are complaints registered at 12 and
13 hours, however, why is not clear from the picture. This is another
illustration of the observation that features other than physical or
chemical ones may be involved in triggering complaints by the
population.

Table IV. Karhuhnen-Loève Transformation of "Schiedam" Data
(Autoscaled and Weighted)

	Eigenvector	
Feature	First	Second
Sulfur dioxide (loc.10)	0.30	0.00
Sulfur dioxide (loc.11)	0.15	0.23
Direction of wind *	0.01	0.49
Variation in sulfur dioxide (loc.11)	0.12	0.04
Sulfur dioxide (loc.12)	0.16	0.09
Sulfur dioxide (loc.13)	0.12	0.!1
Variation in sulfur dioxide (loc.10)	0.12	0.04
Saturated hydrocarbons	0.02	0.00
	1.00	1.00

Eigenvalue (i.e., information) 54% + 14% = 68%

Table V. Karhunen-Loève Transformation of "Schiedam" Data
(Logarithmized, Weighted and Autoscaled)

	Eigenvector	
Feature	First	Second
Log Sulfur dioxide (loc.10)	0.36	0.00
Log Sulfur dioxide (loc.11)	0.28	0.32
Log Sulfur dioxide (loc.12)	0.12	0.21
Log Sulfur dioxide (loc.13)	0.10	0.36
Log Variation SO2 (loc.10)	0.06	0.04
Log Variation SO2 (loc.11)	0.06	0.03
Direction of wind*	0.02	0.04
	1.00	1.00

Eigenvalue (i.e. information) 58% + 13% = 71%

* Integrated backwards in time for 4 hours.

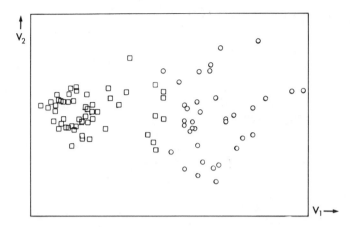

Figure 2. Projection of hours O, with complaints, and □, without complaints, of air pollution on the two most significant eigenvectors of the Karhunen-Loeve transformed, seven-dimensional feature space. Reproduced with permission from Ref. 7. Copyright 1984, The Royal Society of Chemistry.

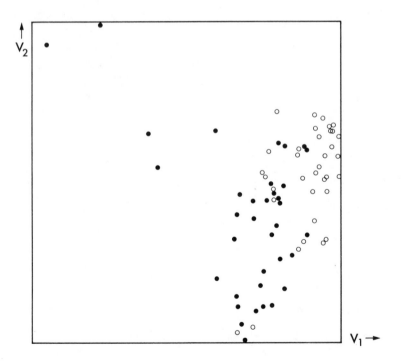

Figure 3. Projection of hours ●, with complaints, and O, without complaints, of air pollution on the two most significant eigenvectors. See Table IV.

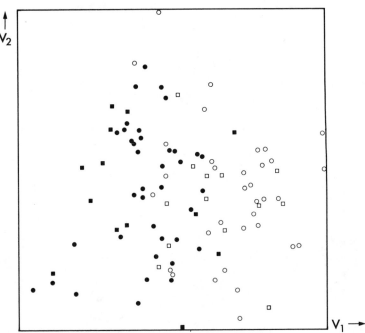

Figure 4. Projection of hours ●, with complaints, and ○, without complaints, including test set patterns ■ , with complaints, and □ , without complaints. See Table V.

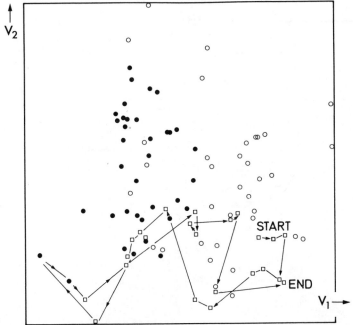

Figure 5. Projection of hours ● , with complaints, and ○ , without complaints, including a test set of 24 hourly measurements on May 17, 1982, ● , with complaints, and □ , without complaints, starting at 0.00 h.

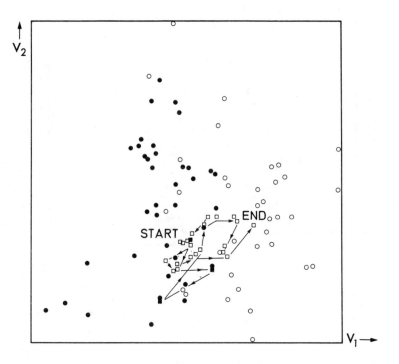

Figure 6. Projection of hours ●, with complaints, and ○, without complaints, including a test set of hourly measurements on July 7 and 8 ■, with complaints, and □ without complaints.

Statistical and mathematical procedures

In order to treat all features without preference, they are scaled such that all feature-axes in the multi-dimensional feature space get an equal length according to

$$x'_{i,k} = (x_{i,k} - \bar{x}_i) / \{ (N-1)^{\frac{1}{2}} \sigma (x_i) \}$$

which is named autoscaling. Here $x'_{i,k}$ represents the autoscaled feature i for pattern k; $\bar{x}_i = \sum_k x_{i,k}/N$; N the number of patterns and $\sigma (x_i)$ the standard deviation of feature i, according to

$$\sigma (x_i) = \{ \sum_{k=1}^{N} (x_{i,k} - \bar{x}_i)^2 /(N-1) \}^{\frac{1}{2}}$$

Thus the center of each axis equals zero and the distribution around the center becomes symmetrical for Gaussian distributed feature values; $\sigma (x_i)$ represents the unit length along the axes.

The distance $d_{i,j}$ between two patterns i and j in the multi-dimensional feature space is calculated according to the Euclidean distance definition:

$$d_{i,j} = \sum_{k=1}^{M} \{ (x'_{k,i})^2 - (x'_{k,j})^2 \}^{\frac{1}{2}}$$

where M represents the number of features and thus the dimensionality of the feature space. $x'_{k,i}$ represents the autoscaled value for feature k with pattern i. The distances are collected in the distance matrix with the dimension N * N. This matrix is symmetrical around the diagonal; all diagonal elements are zero.

The hierarchical clustering procedure operates on the distance matrix. Clustering of patterns is searched for by combining patterns with high similarity into gravity centres, in between the similar patterns. Here, a similarity scale runs from 1 to 0 according to

$$S_{i,j} = 1 - d_{i,j} / D_{max}$$

where D_{max} represents the highest numerical value encountered in the distance matrix. After each combination of two patterns or of a pattern and a gravity center, the distance matrix is recalculated. The procedure is followed graphically and ended when a preset number of clusters is found or when the gravity centers of the clusters upon combination have to travel beyond a preset distance. The graphical representation of the clustering retains distances or similarities but omits the mutual orientation of various patterns.

The minimal spanning tree also operates on the distance matrix. Here, near by patterns are connected with lines in such a way that the sum of the connecting lines is minimal and no closed loops are constructed. Here too the information on distances is retained, but the mutual orientation of patterns is omitted. Both methods, hierarchical clustering and minimal spanning tree, aim for making clusters in the multi-dimensional space visible on a plane.

The correlation between two features p and g reads

$$C_{p,g} = \frac{1}{(N-1) \cdot \sigma_p \cdot \sigma_g} \Sigma^N_{k=1} \left(x_{p,k} - \bar{x}_p \right) \left(x_{g,k} - \bar{x}_g \right)$$

where all symbols have the meanings as defined above.

The weight of a feature depends on its ability to separate two categories or clusters j and k from each other. The equation for the variance weight $W_{i,j,k}$ for feature i reads.

$$W_{i,j,k} = \frac{\overline{(x')^2}_{j,i} + \overline{(x')^2}_{k,i} - 2 \overline{x'_{j,i}} \; \overline{x_{k,i}}}{(M2)_{i,j} + (M2)_{i,k}}$$

Here $x'_{j,i}$ represents the autoscaled value for feature i for a pattern situated in cluster j and (M2) the second moment for all feature values i of the N_j patterns in cluster j according to

$$(M2)_{i,j} = \Sigma^{N_j}_{k=1} \left(x'_{i,k} - \bar{x}'_{i,k} \right)^2 / N_j$$

The Karhuhnen-Loève transformation represents an eigenvalue-eigenvector calculation based upon the variance-covariance matrix of the features. It aims for a linear combination of features such that there are as much linear combinations as features. The linear combinations are mutually orthogonal and have a norm equal to one. Each linear combination (eigenvector) accounts for a part of the total variance encompassed by the features. The vectors are ranked according to this variance - the eigenvalue.

Conclusions

Pattern recognition offers a useful tool for the description of air pollution in industrialized areas. Depending on the weather conditions, sometimes even a prediction of situations with bad-smelling air may be obtained. However, when the weather conditions are unstable, no valid prediction is possible. Apart from physical, meteorological and chemical features, other factors must be accounted for to predict the burden felt by people living in the area.

Acknowledgments

Thanks are due to J.E. Evendijk, J.H.C. Eilers and P.J.W.M. Muskens, D.C.M.R., for making the measured data on air constituents available to us; to G.A.P.E. Jacobs and G.J.H. Roelofs who did the computations and computer programming during their graduate studies at our laboratory; and to B.R. Kowalski for making the computer program "ARTHUR" available to us.

Literature Cited

1. Tan J.T., and Gonzales, R.C., "Pattern recognition Principles," Addison-Wesley, Reading, MA, 1979.
2. Jurs, R.C., and Isenhour, T.L., "Chemical Applications of Pattern Recognition," Wiley, New York, 1975.
3. Isenhour, T.L., Kowalski, B.R., and Jurs, R.C., CRC Crit. Rev. Anal. Chem. 1974, 4, July, 1.
4. Kateman, G., and Pijpers, F.W., "Quality Control in Analytical Chemistry," Wiley, New York, 1981.
5. Massart, D.L., and Kaufman, L., "The Interpretation of Analytical Chemical Data by the Use of Cluster Analysis," Wiley, New York, 1983.
6. Duewer, D.L., Koskinen, J.R., and Kowalski, B.R., "ARTHUR," Laboratory for Chemometrics, Department of Chemistry BG10, University of Washington, Seattle, WA.
7. Pijpers, F.W., Analyst, 109 299 (1984)

RECEIVED July 17, 1985

8

Application of Soft Independent Modeling of Class Analogy Pattern Recognition to Air Pollutant Analytical Data

Donald R. Scott

Environmental Monitoring Systems Laboratory, U.S. Environmental Protection Agency, Research Triangle Park, NC 27711

The SIMCA 3B computer program is a modular, graphics oriented pattern recognition package which can be run on a microcomputer with limited memory, e.g., an Osborne 1 with 64K memory. The SIMCA program was used to display small (four to eight objects) analytical data sets for exploratory data analysis after principal component fitting. K-Nearest Neighbor distances were also computed. The data sets included an interlaboratory comparison of trace element analyses of simulated particulates by X-ray emission; a comparison of flame and Zeeman atomic absorption methods for analyzing lead in gasoline; and GC/MS analysis of volatile organic compounds in ambient air. The combination of principal component and K-Nearest Neighbor analysis was found to provide a convenient and quick method for determining trends and detecting outliers in the data sets.

Pattern recognition has been applied in many forms to various types of chemical data ($\underline{1,2}$). In this paper the use of SIMCA pattern recognition to display data and detect outliers in different types of air pollutant analytical data is illustrated. Pattern recognition is used in the sense of classification of objects into sets with emphasis on graphical representations of data. Basic assumptions which are implied in the use of this method are that objects in a class are similar and that the data examined are somehow related to this similarity.

Before analysis, it is necessary to arrange the relevant data in a data matrix which consists of n objects (laboratories, samples, methods, etc.) arranged in rows with p columns of variables (concentrations, peak heights, etc.). The objects are designated with a subscript i, and the variables are designated with a k. An element in the matrix, X_{ik}, represents the value of variable k for object i. Columns show the values of the particular variable k over all n objects, and rows show the values of all p variables for a particular object i.

Simca Pattern Recognition

The SIMCA pattern recognition techniques were developed by S. Wold and co-workers at the University of Umea, Sweden (3,4). SIMCA is an acronym for Soft Independent Modeling of Class Analogy. A version of these procedures, SIMCA 3B, is now available which will run on a microcomputer. The computer programs are user interactive and graphically oriented. In this study an Osborne 1 microcomputer with a CP/M operating system and 64K memory was used. This amount of memory is sufficient to handle a data matrix of size 50 objects by 50 variables. The compiled program occupied 220K space on a double density floppy disk.

The SIMCA 3B package includes modules to define a data file; to scale, weight, and transform data; to edit, merge, or split the data file; to list the file; to input the data, define classes, and perform K-Nearest Neighbor analysis; to plot the data; to perform principal component analysis for classes; to test the fit of data to defined classes; and to predict values of dependent variables from relationships with independent variables. A flow chart of the various modules in SIMCA 3B and their relationships is shown in Figure 1. The SIMCA pattern recognition programs are based on deriving disjoint principal component models for classification of objects and canonical partial least squares procedures for establishing quantitative relationships among variables. The object of the analysis may be to obtain an overview of the data set, to reduce the number of variables to the most important ones, to determine correlations between variables, to classify objects, or to determine outliers in the data set. The objects in the data matrix may be laboratories, methods, samples analyzed, chemical elements, or compounds depending upon the problem. The variables may be peak intensities, concentrations, etc., but they must be relevant to the problem. Each row of variables in the data matrix must pertain to the same property for all objects. Two assumptions are important when using these procedures. The user must know the type of information desired from the data, and the data must have been derived from relevant, well performed experiments.

Each object in the data table can be considered to represent a single point in a k-fold space (called measurement space) defined by the row vector of k variables listed in the data table. Each of the k variables in the row vector represents the value of the coordinate of the object point along the kth axis in this measurement space. Since the number of variables for a given data set can be very large, the resulting measurement space can have a correspondingly large number of dimensions. This large dimensionality of the data makes it very difficult to obtain an overview of the data set. If the objects are similar with regard to the variables used, then the points in measurement space should form a cluster or class. On the other hand outliers in the data set should be noticeable by their distance from this cluster or class. The definition of the class or classes of objects will depend upon the number of objects in the class and the relevance of the variables used for the objects.

Since similar objects should be relatively close to each other
in measurement space, one method of classifying objects is by their
distances from each other in this space. These distances are calculated in the K-Nearest Neighbor Module, included in the SIMCA package.
The distance, d, between objects j and 1 is determined from the
Euclidean distance:

$$d^2_{j1} = \sum_k (X_{jk} - X_{1k})^2$$

An object can be classified by the distances to its nearest neighbors, e.g., the nearest three.

One of the important functions of principal component analysis is
the reduction of dimensionality so that an overview or graphical
representation of the data set may be given in two dimensional plots.
This allows the user to "see" the relationships between the objects
in the data set. This process is accomplished by fitting two or more
principal components to the data. The first component is oriented
along the axis of greatest variance of the variables in the data
matrix about their means. The second principal component is independent of (orthogonal to) the first and is the vector along the axis of
next greatest variance in the data. Succeeding principal components
can be calculated which will be orthogonal to the preceding ones and
which may explain some of the remaining variance. The principal
components are linear combinations of the original variables which are
fitted in the least squares sense through the points in measurement
space. These new variables usually result in a reduction of variables from the original set and often can be correlated with physical
or chemical factors. The coefficients of the original variables in
the principal components, the loadings, provide information regarding
important and redundant variables for the analyzed data. SIMCA uses
a bilinear projection model to decompose the data matrix into a score
matrix, a loading matrix, and a residual matrix. The residual matrix
contains that part of the data matrix which does not fit the model.
If the residuals are small compared with the variation in the data
matrix, then the model is a good representation of the data matrix.

In the following discussion, three types of air pollutant analytical data will be examined using principal component analysis
and the K-Nearest Neighbor (KNN) procedure. A set of interlaboratory
comparison data from X-ray emission trace element analysis, data from
a comparison of two methods for determining lead in gasoline, and
results from gas chromatography/mass spectrometry analysis for volatile organic compounds in ambient air will be used as illustrations.

X-ray Emission Methods for Trace Element Analyses

An intercomparison study of trace element determinations in simulated and real air particulate samples has been published by Camp,
Van Lehn, Rhodes, and Pradzynski (5). This involved twenty-two
different laboratories reporting up to thirteen elements per sample.
The simulated samples consisted of dried solution deposits of ten
elements on Millipore cellulose membrane filters. In our data analysis a set of energy dispersive X-ray emission results restricted to
eight laboratories reporting six elements (V, Cr, Mn, Fe, Zn, Cd) was

used for the simulated samples. The data table for the simulated
samples is shown in Table I, which is an eight laboratory (objects)
by six element (variables) matrix. The cadmium data for laboratories
2 and 7 were not reported but were estimated to be the median of the
remaining elements for each laboratory. All of these data have been
scaled by dividing the reported values by the known ones. Therefore, a value of 1.00 represents a perfect analysis result. The
median over all elements for each laboratory is also given in the
Table. These range from a low of 0.76 for Laboratory 8 to a high of
1.02 for Laboratory 4. Laboratories 4 and 8 are suspected outliers
in this set of data.

Table I.   Interlaboratory Comparison of X-ray Emission
Analyses, Simulated Samples[a]

Lab	V	Cr	Mn	Fe	Zn	Cd	Median[b]
1	0.73	0.84	0.80	0.86	0.64	1.01	0.82
2	0.87	0.90	0.88	0.87	0.90	(0.88)[d]	0.88
3	1.01	0.87	0.92	1.00	0.94	1.09	0.97
4[c]	1.03	1.01	1.00	1.11	1.44	0.85	1.02
5	0.97	0.95	0.96	0.95	0.95	0.94	0.95
6	0.94	0.95	0.94	0.94	1.03	0.97	0.95
7	0.99	0.98	0.96	0.92	0.91	(0.95)[d]	0.96
8[c]	0.71	0.77	0.75	0.69	0.82	0.89	0.76

[a]Reference 5. All data have been scaled by dividing the reported
values by the known ones.
[b]Median of each laboratory results, omitting estimated data.
[c]Suspected outliers.
[d]Data not reported. Estimated as laboratory median for all elements.

A principal component analysis was performed on all the data in
Table I after autoscaling the data. The results of the analysis are
given in Table II. The scaled averages and the weights for each
variable are listed. The loadings and modeling power for each variable also are listed with the remaining unexplained variance in the
data table and the residual standard deviation for each principal
component. The modeling power is the percentage of the standard
deviation for the variable which is explained by the principal component. The residual standard deviation is the standard deviation of
the data matrix which is not explained by the model. From the loadings in Table II it can be seen that the first principal component is
composed of approximately equal weights of all elements except cadmium
which has a zero modeling power and a low negative loading. The
first and second principal components together account for 89% of the
original variance with a residual standard deviation of 0.40. The
loadings of the second principal component show a very strong contribution from cadmium with smaller ones from zinc, vanadium, and iron.
The iron and vanadium loadings are essentially equal in both principal
components. All of the variables, except cadmium, are modeled well

by both principal components. A plot of the locations of the eight laboratories on the two principal components is shown in Figure 2. This is a projection of the objects from six variable space onto the plane defined by the two principal components. The laboratories are scattered with 4, 8, and 3 apparently on the fringes of the data set. Examination of the loadings of the variables for the two principal components shows that cadmium is widely separated from the other clustered elements, indicating that it is anomalous. Therefore, the cadmium data were omitted from the data matrix, and the principal component analysis was performed again. The results show that Laboratories 4 and 8 are indeed outliers with the other laboratories clustered together.

Table II. PC Analysis of X-ray Data, Simulated Samples

Elements	V	Cr	Mn	Fe	Zn	Cd
Average[a]	7.256	11.425	10.448	7.573	4.185	12.210
Weights[b]	8.007	12.572	11.593	8.254	4.388	12.887
			PC1			
Loading	0.452	0.451	0.471	0.445	0.408	-0.072
Model Power[c]	0.63	0.62	0.78	0.59	0.43	0
Remaining Variance[d]:			32.8%			
Residual Std. Deviation[e]:			0.627			
			PC2			
Loading	-0.216	0.070	-0.097	-0.200	0.337	-0.886
Model Power	0.71	0.60	0.79	0.63	0.56	0.86
Remaining Variance:			10.8%			
Residual Std. Deviation:			0.402			

[a]Scaled averages obtained by multiplying original averages by weight.
[b]Weights are the reciprocal of the standard deviation for each variable.
[c]Percentage of explained standard deviation for this variable after the respective PC fit.
[d]Variance remaining after principal component fit.
[e]Standard deviation of data matrix unexplained by PC model.

In order to check the results of the analysis, K-Nearest Neighbor distances were computed for the scaled data set including the cadmium results. The median of the distances from a given laboratory to the three nearest neighbors ranged from 0.26 to 1.24 with the median distance between members of the cluster (1,2,3,5,6,7) equal to 0.79. The median distances of Laboratories 4 and 8 from members of this cluster were 1.24 and 1.22, respectively, supporting the view that these laboratories are outliers.

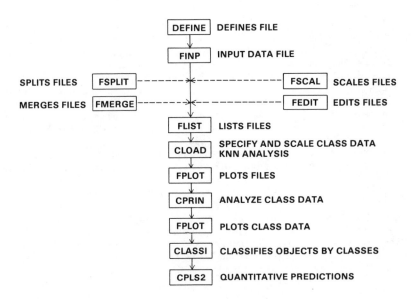

Figure 1. SIMCA module flowchart.

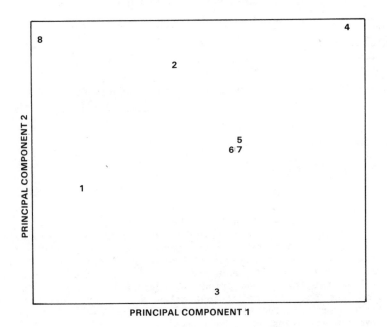

PRINCIPAL COMPONENT 1

Figure 2. Principal component plot of x-ray data by laboratory.

Methods of Determining Lead In Gasoline

The ASTM-EPA standard method of analyzing lead in gasoline requires extraction of alkyl lead iodide complexes into methylisobutylketone and a subsequent flame atomic absorption analysis of the extract. A more direct method has been proposed (6) which uses Zeeman atomic absorption analysis after sample dilution. Both methods were used to analyze a set of five field collected samples. The results showed a bias (average difference between method results) of 0.0012 g/gal with the standard flame results higher. The correlation coefficient between the results was 0.9998 ± 0.0009, and a pairwise t-test showed no difference between the methods (6).

This very small set of results was subjected to principal component and K-Nearest Neighbor analysis, after autoscaling, to determine the effectiveness of the SIMCA procedures in displaying small data sets. The data table (four methods by five samples) is shown in Table III. The methods to be compared are two of the objects, and two other objects were artificially constructed by adding and subtracting 0.01 g/gal to the standard flame atomic absorption results. The ± 0.01 g/gal range is an Environmental Protection Agency guideline for replicate analysis results within one laboratory. This range can be used to gauge the spread of the objects in the principal component plot.

Table III. Comparison of Lead in Gasoline Methods

Sample Method	1	2	3 (g/gal)	4	5
Flame AAS [a]	0.0015	0.0040	0.0040	0.0105	0.082
Zeeman AAS	0.0015	0.0046	0.0039	0.0110	0.075
Flame +.01 [b]	0.0115	0.0140	0.0140	0.0205	0.092
Flame -.01 [b]	-0.0085	-0.0060	-0.0060	0.0005	0.072

Source: Adapted from Reference 6.
[a] The ASTM-EPA standard method.
[b] Data constructed by adding or subtracting 0.01 g/gal from the standard method result.

The results of the principal component analysis are given in Table IV and Figure 3. The first principal component is composed of essentially equal contributions from all samples with very good modeling for all samples. After the fit of the second principal component, there was no remaining variance and no residual standard deviation. The loadings for the second principal component showed a high contribution from the highest concentration sample with approximately equal but lower negative contributions from the rest. The modeling power for all variables was excellent.

Table IV. PC Analysis of Lead in Gasoline Data

Samples	1	2	3	4	5
Average	0.1837	0.5079	0.4868	1.301	9.034
Weights	122.5	122.4	122.5	122.4	112.6
			PC1		
Loading	−0.453	−0.451	−0.453	−0.452	−0.427
Model Power	0.92	0.88	0.93	0.88	0.59
Remaining Variance:	4.1%				
Residual Std. Dev.:	0.228				
			PC2		
Loading	−0.172	−0.270	−0.155	−0.254	0.900
Model Power	1.00	1.00	1.00	1.00	1.00
Remaining Variance:	0%				
Residual Std. Dev.:	0.000				

Examination of the projection plot in Figure 3 gives the impression that the new Zeeman method (ZAAS) is farther from the standard flame method (FAAS) than are the guideline limits of ± 0.01 g/gal. To verify this impression, a K-Nearest Neighbor distance analysis of the scaled data was conducted. The resulting distances are shown on the vectors in Figure 3. The distance between the standard method and the Zeeman method is 0.355 compared to a distance of 1.21 from the standard method to either guideline limit. These results confirm the equivalence of the results from the two methods as already noted using parametric statistical methods (6). They also indicate a problem with visual examination of principal component plots of data sets as small as the present one. The use of the K-Nearest Neighbor distances provides a convenient check on the principal component plots in this case.

Gas Chromatography–Mass Spectrometric Analysis of Organic Compounds

A commonly used method of sampling and analysis for volatile organic compounds in ambient air is by concentration of the compounds on a solid sorbent such as Tenax and subsequent thermal desorption and GC/MS analysis of the collected compounds. The analysis phase, although not trivial, can be done well if proper care is taken. However, the sampling phase of this process apparently introduces artifacts and unusual results due to, as yet, unknown factors. A method to detect some sampling problems has been proposed and tested (7). This distributed air volume method requires a set of samples of different air volumes to be collected at different flow rates over the same time period at the sampling location. Each pollutant concentration for the samples should be equal within experimental error since the same parcel of air is sampled in each case. Differences in results for the same pollutant in the various samples indicates sampling problems.

A set of results from the distributed air volume method obtained in Elizabeth, New Jersey, in October, 1981, (7) is given in Table V. There are four objects (samples with 10, 15, 21, and 26 L total air volume) with eight organic compounds as variables. A cursory examination of the data table shows that the results from the 10 L sample are higher than those for the other samples for all pollutants. Thus this sample is a suspected outlier.

Table V.   GC/MS   Data   for   Organic   Compounds   Collected   on   Tenax

Compound Sample	1	2	3	4 ($\mu g/m^3$)	5	6	7	8
10 L[a]	8.64	18.63	3.74	5.10	1.86	1.48	4.69	11.92
15 L	7.12	15.04	3.01	2.81	0.81	1.37	2.69	5.29
21 L	6.17	14.89	3.11	3.44	1.11	1.58	2.99	5.00
26 L	6.76	15.05	2.96	2.68	0.73	1.69	2.59	5.04

Source: Adapted from Ref. 7.
                    Compounds: 1(benzene),   2(toluene),   3(1,2-dimethylbenzene),   4(ethylbenzene),   5(styrene),   6(trichloroethylene),   7(1,1,1-trichloroethane),   8(benzaldehyde).
[a]Suspected outlier.

The autoscaled data were subjected to a principal component analysis with the results given in Table VI. The first principal component was composed of approximately equal contributions from all compounds except for trichloroethylene, which had a low contribution. The results for modeling power for the first principal component showed a good fit for benzene and a very good fit for all the other compounds except trichloroethylene, which had zero modeling power. The fit of the first and second principal components gave a remaining variance of 11% with a residual standard deviation of 0.38. The second principal component was composed of trichloroethylene with very small contributions from the other compounds. Benzene had a negative and low contribution to this principal component. The modeling power was very good for all compounds except benzene. The projection of the samples on the plane of the two principal components is shown in Figure 4 where it appears that the 10 and 15 L samples are not close to the 21 and 26 L samples. The loadings of the two principal components show that six of the compounds clustered, with benzene and trichloroethylene outside the cluster. To confirm the results of the principal component projections, a K-Nearest Neighbor analysis of the scaled data was performed. The 10 L sample had a median distance from the other samples of 1.84. The median distance between the members of the (15, 21, 26 L) cluster is 0.71. This supports the identification of the 10 L sample as an outlier. However, the 15 L sample does not appear to be an outlier.

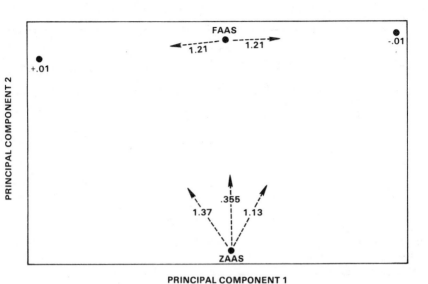

Figure 3. Principal component plot of lead methods and KNN results.

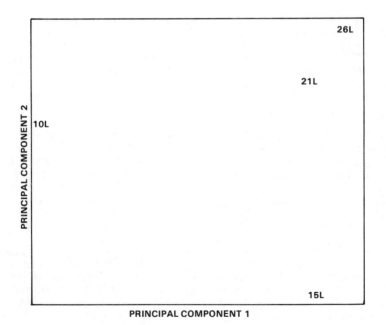

Figure 4. Principal component plot of GC/MS data by sample.

Table VI. PC Analysis of GC/MS Data

Compound	1	2	3	4	5	6	7	8
Average	6.81	8.74	8.85	3.15	2.19	11.2	3.30	2.00
Weights	0.95	0.55	2.76	0.90	1.94	7.3	1.01	0.29

PC1

Loading	0.35	0.38	0.38	0.37	0.37	-0.12	0.38	0.38
Model Power	0.49	0.82	0.88	0.72	0.70	0	0.86	0.86
Remaining Variance:		23%						
Residual Std. Dev.:		0.52						

PC2

Loading	-0.18	0.06	0.07	0.12	0.11	0.96	0.09	0.03
Model Power	0.36	0.76	0.87	0.66	0.62	0.79	0.87	0.81
Remaining Variance:		11%						
Residual Std. Dev.:		0.38						

Compounds: 1(benzene), 2(toluene), 3(1,2-dimethylbenzene), 4(ethyl-benzene), 5(styrene), 6(trichloroethylene), 7(1,1,1-trichloroethane), 8(benzaldehyde).

After omitting the 10 L data, the data were resubjected to a principal component analysis. The loadings of the two new principal components were compared with those from the previous analysis. The omission of the 10 L data caused a separation of benzaldehyde and toluene from the previously clustered compounds as well as retaining the separation of benzene and trichloroethylene as found previously.

Conclusions

The use of the SIMCA principal component analysis and graphing pro-grams to obtain a "window" into the multi-dimensional measurement space of a data set is a quick and effective way to obtain an over-view of a data set and to detect outliers. For very small numbers of objects, e.g., the four in the data from the lead in gasoline methods comparison or the Tenax GC/MS data, it is necessary to confirm the graphical results by performing K-Nearest Neighbor analysis of the data. The analysis of the data from the lead in gasoline methods comparison also showed that useful results can be obtained even with only two real objects (the two methods) with five variables (samples). Finally, it is best to remember what S. Wold has said about statis-tical methods of data analysis (3), "In a data set there is often no information whatsoever about the given problem." Therefore, par-ticular care should be taken to design experiments to answer the desired questions and to guard against over-analysis of irrelevant or noisy data.

Disclaimer

This paper has been reviewed in accordance with the U. S. Environmental Protection Agency's peer and administrative review policies and approved for presentation and publication. Mention of trade names or commercial products does not constitute endorsement or recommendation for use.

Literature Cited

1. Kowalski, B. R. "Chemometrics: Mathematics and Statistics in Chemistry"; D. Reidel Publishing Co.: Boston, 1984.
2. Kowalski, B. R. "Chemometrics: Theory and Application"; ACS SYMPOSIUM SERIES No. 52, American Chemical Society: Washington, D.C., 1977.
3. Wold, S., et al. In "Chemometrics: Mathematics and Statistics in Chemistry"; Kowalski, B. R., Ed.; D. Reidel Publishing Co.: Boston, 1984; p.17.
4. Wold, S., et al. In "Food Research and Data Analysis"; Martens, H.; Russwurm, H., Eds; Applied Science Publishers: New York, 1983.
5. Camp, D. C.; Van Lehn, A. L.; Rhodes, J. R.; Pradzynski, A. H. X-Ray Spectrometry 1975, 4, 123.
6. Scott, D. R.; Holboke, L. E.; Hadeishi, T. Anal. Chem. 1983, 55, 2006.
7. Walling, J. Atmos. Environ. 1984, 18, 855.

RECEIVED June 28, 1985

9

Cluster Analysis of Chemical Compositions of Individual Atmospheric Particles Data

T. W. Shattuck[1,2], M. S. Germani[2], and P. R. Buseck[2]

[1]Department of Chemistry, Colby College, Waterville, ME 04901
[2]Departments of Chemistry and Geology, Arizona State University, Tempe, AZ 85287

Atmospheric particle types are identified using k-means cluster analysis. Nearest neighbor classification is used to produce particle number versus type histograms that allow identification of spatial and temporal emission patterns. Factor analysis is carried out on the particle-type results from several sampling periods or sites to identify relationships between particle types and for source identification. The methods are applied to the elemental composition of particles from the Phoenix aerosol which are obtained using an automated analytical scanning electron microscope. Seven methods are considered for choosing cluster seedpoints. Cluster significance is judged using the ratio of the sum of squared distances between clusters to the sum of squared distances within clusters. In order to account for the full variability in the data set, more clusters are necessary than may be statistically significant.

Data obtained from the analysis of individual atmospheric particles is ideal for the identification of particle sources and for the study of particle dynamics and emission patterns (1). Using an analytical scanning electron microscope (ASEM) equipped for energy-dispersive X-ray spectrometry (EDS), the elemental composition, size, shape and morphology of particles can be determined. This information is necessary for determining the effects of particles on such important areas as health, climate and visibility. Individual particle analysis is particularly useful for studying elemental speciation and association, particle agglomeration, surface coatings and the distribution of elements as a function of particle size (2-6). The ASEM in our laboratory is automated so that analyses of about 1000 particles are commonly used to characterize each sample. The ability to rapidly analyze large numbers of particles necessitates the development of statistical methods for data reduction and analysis of these large data sets.

0097-6156/85/0292-0118$06.00/0

Cluster analysis is used to determine the particle types that occur in an aerosol. These types are used to classify the particles in samples collected from various locations and sampling periods. The results of the sample classifications, together with meteorological data and bulk analytical data from methods such as instrumental neutron activation analysis (INAA), are used to study emission patterns and to screen samples for further study. The classification results are used in factor analysis to characterize spatial and temporal structure and to aid in source attribution. The classification results are also used in mass balance comparisons between ASEM and bulk chemical analyses. Such comparisons allow the combined use of the detailed characterizations of the individual-particle analyses and the trace-element capability of bulk analytical methods.

These methods, while being developed for the study of the Phoenix aerosol, are also applicable to a wide range of studies. The vast majority of particles >1um in diameter in the Phoenix aerosol are crustal in origin, representing a wide variety of mineral particles. They thus provide a stringent test case for the methods, since these particles produce many large, closely spaced clusters, and these tend to obscure smaller, atypical clusters that are of anthropogenic origin.

Cluster Analysis

There are three goals for cluster analysis. 1) The most immediate is the qualitative identification of the types of particles that occur in an aerosol. The compositions of the clusters often directly indicate sources. For example, particles containing Pb, Cl and Br indicate auto exhaust. The clusters may also provide information on formation mechanisms. For example, a cluster composed mostly of calcium and sulfur but with a small amount of silicon and a few percent of transition metals suggests a $CaSO_4$ particle with a silicate core which is most likely formed as a result of combustion processes. 2) The next goal is to reduce the mass of data to a tractable size, but in a way that emission patterns can be easily discerned. This is done by using the cluster centroids from representative samples to define the particle types in the aerosol. Particles from the remainder of the data set are assigned to the various particle types. Histograms of the number of particles for each particle type, for each sampling site and period, provide a rapid way to follow temporal and spatial emission patterns. The particle type classifications also are used as input for factor analysis. 3) The third goal is to allow poorly populated clusters to be treated separately from the clusters containing many particles. An example of the need for this separation arises in the Phoenix aerosol. This is because about 75% of the particles >1.0 um in diameter in the Phoenix aerosol are quartz or alumino-silicate mineral particles which make it difficult to monitor particles of similar size that are not of crustal origin. Particles that are not represented by a cluster are left unassigned. These unassigned particles are particularly useful for studying unusual events. However, this requires that the cluster analysis is sufficiently inclusive so that only unusual particles are in the set of unassigned particles. Such separation is particularly important if

there is a subsequent need to return to those particles for further analysis.
There are three steps to nonhierarchical cluster analysis. The first is to choose seedpoints; these are approximate points compositions from which to start cluster analysis. Choosing seedpoints is by far the most critical step. Secondly, a cluster-analysis algorithm is applied to define the clusters. Finally, the statistical significance of the clusters must be determined. In other words, are the clusters well resolved or do they overlap? The three steps are detailed below.

Choosing Seedpoints. A group of successive observations or a set of observations chosen at random from the data set may be used for seedpoints. However, the results of such simple procedures are often not reliable. Seven different methods are considered for choosing seedpoints in this study. The first four are standard hierarchical techniques; single, complete, average (between merged groups) linkage and Ward's method (7). Nearest centrotype sorting (8) is used to choose seedpoints, which, in common with Ward's method, seeks to minimize the sum of squared distances as an objective criterion. The first refinement method, here called the "refine" procedure, from the FASTCLUS procedure in the SAS statistics package (9),is also considered as a seedpoint technique. Finally, a technique that will be called the "merge" procedure completes the seedpoint location techniques. The latter three methods will be described below. In each case, the general procedure starts by choosing a number of successive observations from the data set. This initial set is reduced to the final desired number of seedpoints by one of the seven methods. Euclidian distances are used.
Nearest centrotype sorting uses the assumption that observations as seedpoints that minimize the sum of within-cluster distances best represent the clusters (8). This is illustrated in Figure 1. The observation chosen as a seedpoint in Figure 1a is a much better choice because it gives a smaller sum of distances (and squared distances). In the sorting procedure, each successive observation chosen as a seedpoint is the observation that produces the greatest reduction in the total sum of distances. Nearest centrotype sorting, while an excellent method, is slow and so is not particularly useful in an interactive mode. It is considered here mainly for comparison to the other methods.

The "refine" procedure starts by assigning a trial set of observations as seedpoints; it then tests the remaining observations to see if they are better seedpoints than those already chosen. The test is carried out as follows. The distance from the observation to be tested to the closest seedpoint is found, d_t. This is compared to the distance between the closest two seedpoints, d_c. If $d_t < d_c$, the seedpoint set remains unchanged and another observation is tested. If $d_t > d_c$, then the test observation is chosen as a seedpoint, and the seedpoint from the closest pair nearer the test observation is rejected . For example, in Figure 2a the observation to be tested is shown as an open circle, and those observations that have previously been assigned as seedpoints are shown as solid

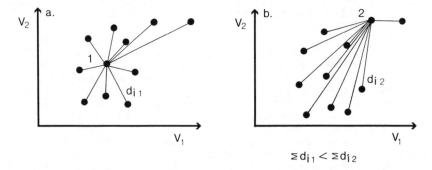

$$\Sigma d_{i1} < \Sigma d_{i2}$$

Figure 1. Nearest centrotype sorting minimizes the sum of distances of each observation to its closest centrotype as in (a).

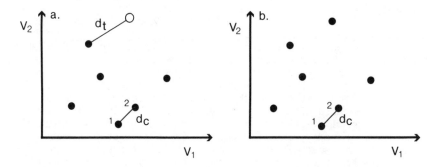

Figure 2. a) "Refine" procedure, b) "merge" procedure for choosing seedpoints. In 2a) the observation represented by the open circles would be accepted as a seedpoint. In both cases observation 2 would be rejected.

circles. The distances to be compared are labeled. In this case
$d_t > d_c$, so the test observation is chosen as a seedpoint and
seedpoint 2 is rejected.

In the "merge" procedure the initial set of observations is
reduced by scanning the set repeatedly, each time locating the two
observations that are closest to one another and rejecting the
second of the two. This rejection is repeated until the final
desired number of seedpoints remains. For example, in Figure 2b the
data points are the initial set of observations from the full data
set. The closest observations are indicated, and the second of them
is to be rejected.

The "merge" procedure is in many ways the reverse of the
commonly used "simple cluster seeking" method of Tou and Gonzalez
(10). In their method a minimum seedpoint separation distance is
specified. The first observation is chosen as a seedpoint. The
next observation that is greater than this minimum distance away
from the first is also chosen. The rest of the data set is then
scanned for observations that are greater than the minimum distance
from the seedpoints that have already been chosen. The advantage of
the "merge" procedure is that the minimum distance need not be
specified; the choice of the minimum distance can be very difficult.

The "merge" procedure is chosen as the principal seedpoint
selection method for this study because it most simply and rapidly
carries out the goals listed above. In general, several methods
should be used to ensure that the seedpoint set includes all
important clusters.

The "merge", "refine" and single linkage procedures are
particularly good at finding low abundance clusters of atypical
compositions. This property satisfies goal 3, above. But because
of this, these procedures should be used with caution. This is
especially true when the number of clusters is uncertain, which is
the normal case. Seedpoints may be located in low-abundance,
atypical clusters while some closely spaced but well resolved large
clusters may be ignored. This may or may not be the desired result.

The number of observations that can be tested as seedpoints is
limited by the size of the initial set of observations chosen from
the data. Because of time constraints, less than 10% of the data
set is commonly included in the initial set. To avoid this
limitation, the seedpoints are chosen in a two-step process. A
trial set of seedpoints is found in the first round and then used
for cluster analysis. The unassigned observations from the first
round are then sampled for an additional set of seedpoints. The two
sets are then combined. This allows many more of the observations
in the data set to be sampled as possible seedpoints. In each of
the two rounds the seedpoints can be chosen by one of the seven
methods listed above.

Cluster Algorithm. The Forgy variety of k-means cluster analysis
(7) is chosen because of its speed for large data sets. Forgy k-
means cluster analysis is an iterative process. In the first
iteration observations are assigned to the nearest centroid. This
defines the initial clusters. The composition of the observations in
each cluster are then averaged to find approximate centroids. Let
$\bar{x}_k$ be the centroid vector for cluster k, with components $\bar{x}_{kj}$, for

all variables j. Then the average is given by

$$\bar{x}_{kj} = \frac{\sum_{i=1}^{n_k} x_{ij}^k}{n_k}$$

for the n_k observations, x_{ij}^k, in the cluster. If the initial seedpoints are far from the true centroids, then the true and approximate centroids so calculated may not be very close. This may be improved through successive iterations by using the approximate centroids as seedpoints and then repeating the assignment and averaging steps. This continues until the centroids no longer change on subsequent iterations. Cluster centroids are updated at the end of each assignment cycle. The Euclidian distance measure is used. Outliers are excluded by choosing a maximum distance for cluster assignment. Convergence of the centroids may take as many as five iterations of the k-means procedure.

<u>Cluster Significance</u>

There are two goals for significance testing. The first is to estimate the number of clusters in the data and the second is to identify the amount of overlap between the various clusters. Unfortunately, no completely satisfactory statistical test exists. One is faced with a difficult decision, either to ignore the problem or to make do with available testing methods. The simplest, most straight-forward test is chosen for this study, the sum of squares ratio test. Even though the test method may be flawed, it is necessary to underscore the importance and usefulness of statistical measures of cluster separation.

The sum of squares ratio test compares two clusters by finding the ratio of the between-clusters sum of squares (B) to the within-clusters sum of squares(W). This is based on the well known sum of squares decomposition,

$$T = B + W$$

where T is the total sum of squares for the two clusters. For each cluster k with n_k members and centroid vector $\bar{x}_k$,

$$T = \sum_{k=1}^{2} \sum_{i=1}^{n_k} (x_i^k - \bar{x})'(x_i^k - \bar{x})$$

where x_i^k is observation vector i from cluster k and $\bar{x}$ is the mean vector over all the observations in the data set. The prime indicates vector transposition. Then

$$B = n_1(\bar{x}_1 - \bar{x})^2 + n_2(\bar{x}_2 - \bar{x})^2$$

and

$$W = \sum_{k=1}^{2} \sum_{i=1}^{n_k} (x_i^k - \bar{x}_k)'(x_i^k - \bar{x}_k) = \sum_{k=1}^{2} \sum_{i=1}^{n_k} d_{ik}^2$$

The term d_{ik} is the Euclidian distance between observation i and centroid k. The test statistic is then

$$C = B/W$$

The C-statistic is used to test the significance of pairs of clusters under the null hypothesis that the observations are a sample from a single normal population. Hartigan ([11]) and Engleman and Hartigan ([12]) compiled a set of percentage points for C for clustering on one variable, assuming a normal distribution of the observations for optimal clustering obtained by maximizing B/W. These percentage points cannot strictly be used to test the significance of clusters in this study since a) the clustering occurs over many variables (dimensions), b) the clusters obtained are usually at best only locally optimal and c) the underlying observations are not normally distributed. However, applying the C-statistic in a simulation study, using k-means clustering of synthetic data over 3 to 9 variables generated using a rectangular distribution, shows the Engleman and Hartigan percentage points to be useful. The percentage points seem to be rather insensitive to the number of variables. A low confidence level (50%) is normally chosen when applying the percentage points to actual data.

Regardless of the failings of a given statistical test, it is the philosophy of the use of the test that is most important. This is especially clear when addressing the problem of estimating the number of clusters in the data set. In some standard statistical packages this is normally handled in the following way. Cluster analysis is carried out by intentionally using too many seedpoints. The distance between the resulting centroids or the variance of the variables in each cluster is then used to decide which clusters to combine and which clusters to split. Using the intercentroid distance as a criterion has the danger of combining two well resolved but closely spaced clusters. Using the variance as a criterion has the danger of arbitrarily dividing a single large cluster. However, using the sum of squares ratio, as in this study, is a more reliable criterion because it takes into account both the between-centroid distance and the dispersion of the clusters.

In this study, we purposely started with too many seedpoints. The number of seedpoints for analysis and the final seedpoint set is determined in the following way. After an initial round of cluster analysis, the seedpoint which gives the largest number of test failures is rejected. After a seedpoint is rejected cluster analysis is repeated. This process continues until the number of unassigned particles begins to increase rapidly and the number of significant clusters decreases. In our case, it is quite likely that there are several clusters in the final set that are not significant (especially in the group of alumino-silicate clusters), but it is necessary to keep some of them in order adequately to describe the variations that occur when the centroids are used in discriminant analysis for other sampling sites and periods. This situation arises in part because k-means analysis works best on spherical clusters, but many clusters are certainly not spherical. In addition, the natural variability of aerosol particles undoubtedly produces significant overlap between clusters. Since there is, at present, no adequate statistical test for significance and no rapid method for clustering non-spherical clusters, the actual use of the cluster centroids for the classification of particles must serve as the test for the adequacy of the cluster analysis. That is, the usefulness and validity of the results is the ultimate test of the cluster analysis.

Particle Classification

Particle classification is carried out using a nearest neighbor criterion with Euclidian distance. Histograms of the size distribution within each particle type can be generated in addition to particle number versus particle type histograms. Particles are not classified if they are further than a chosen maximum distance from the nearest centroid. Histograms of the distribution of elements in the unassigned particles are useful for following unusual events. (Linear discriminant analysis is not used because of the extreme inhomogeneity of the cluster variance–covariance matrices in the data.)

A particularly powerful use of the classification results is in factor analysis. This will help to uncover interrelationships among the particle types and will provide additional information for source attribution. The results of the factor analysis are also helpful for judging the significance of the cluster analysis, in that if the occupations of two similar particle types are uncorrelated over several samples then this indicates that the particle types and the clusters from which they are derived are significantly different.

Experimental Methods

The elemental compostion of the individual particles used in this study were determined by energy-dispersive, X-ray spectrometry (EDS). The data were acquired using an automated analytical scanning electron microscope (JEOL JSM-35). The automation system includes both sample stage and electron-beam automation, allowing unattended operation. Elemental compositions were obtained from the particle X-ray spectrum by integration of the background-corrected X-ray peak in a region-of-interest about one of the characteristic X-ray lines for each element. The region of interest integrals are converted to relative abundance concentrations by dividing the integral for each element by the sum over all the elements detected in the particle. No other variable normalization was used in order to avoid the inclusion of noise in the form of analytical uncertainty due to the relatively large detection limits inherent in EDS analysis.

Data for 31 elements can be rapidly determined but result in some interferences between elements. No spectral curve fitting or matrix (ZAF) correction schemes were used in this survey study. ZAF correction does not seem to markedly aid the cluster analysis process, presumably because the natural dispersion of the clusters is so large and due to the errors in applying thick film ZAF corrections to small particles. The elements used in this study were Na, Mg, Al, Si, Fe, K, Ca, S, P, Cl, Ti, Mn, Cu, Zn, Cr, Ni, As, Br and Pb. The particles for this study were collected on Nuclepore filters. Particles in the size range of 1 to 15 um in diameter were analyzed.

Cluster analysis is far from an automatic technique; each stage of the process requires many decisions and therefore close supervision by the analyst. It is imperative that the procedure be as interactive as possible. Therefore, for this study, a menu-driven interactive statistical package was written for PDP-11 and VAX (VMS and UNIX) series computers, which includes adequate computer graphics capabilities. The graphical output includes a variety of histograms and scatter plots based on the raw data or on the results of principal-components analysis or canonical-variates analysis (14). Hierarchical cluster trees are also available. All of the methods mentioned in this study were included as an integral part of the package.

Results

The seven seedpoint methods were tested using a data set containing 1000 particles from a representative aerosol sample collected in downtown Phoenix. The first 70 successive observations were chosen from the data as the initial set for choosing seedpoints for each method. Each of the seven methods was applied to reduce this set to 30 seedpoints. The 30 seedpoints were then used in k-means cluster analysis. No two of the seedpoint sets were identical; however, 25 out of 30 final clusters were found in each set. The unique seedpoints were found to be statistically insignificant, and in general, the different methods seemed to be dividing large complex clusters in slightly different ways. Single linkage gave the most unusual set of seedpoints and would seem to be an excellent companion method to the "merge" procedure, especially since it gave an unusually small total number of test failures. However, for general use single linkage does not do a good enough job on clusters with typical composition, such as the alumino-silicate clusters. Ward's method gave a slightly smaller number of test failures than complete and average linkage, but otherwise all three gave comparable results. Nearest centrotype sorting also gave comparable results with this data set. Its use is probably warranted only for clustering data containing similar, closely spaced clusters with few atypical clusters. The "refine" method gave the largest number of pairwise test failures. The "merge" procedure also gave a relatively large number of test failures, but the seedpoints were well balanced between the clusters of typical and atypical composition. All of the methods gave only 2 clusters that contain no significant test failures, except "refine" which gave only one. However, all seven methods gave the same number of clusters with less than 4 test failures. The differences between the methods would have been more pronounced if the final number of seedpoints had been smaller. The "merge" method was used in all of the following studies, with the two-round procedure, described above, for choosing seedpoints. The sum-of-squares ratio test was used to eliminate some of the nonsignificant clusters.

These methods, when applied to the downtown Phoenix aerosol sample, produced a satisfying range of particle types and left unassigned only about 4% of the particles (Table I). The major particle type was quartz which accounted for 19% of the particles. Various alumino-silicate types were the next most abundant. Easily identifiable types included clusters rich in only one to three elements, including iron (7%), calcium (3%), calcium-silicon-iron (4%), calcium-sulfur (1%), lead (3%), lead-chloride-bromide (3%) and titanium (2%). The abundances of these particle types, indicated in parentheses, vary widely from site to site. Many particles rich in heavy metals were found in the unassigned group at this point.

Table I. Cluster Composition for Representative
Phoenix Aerosol Sample

Elemental Composition[a]	Similar Mineral[b]	% Abundance
Si K Al Fe	Orthoclase	7
Si Al K Fe	Muscovite	15
Si Al Fe Ca	Albite/Montmorillonite	14
Si Ca Fe Al	(Epidote)	6
Si Fe Al K	Biotite	4
Si	Quartz	19
Fe Si Al Mg	Ripidolite/Chlorite	2
Fe	Magnetite	7
Ca Si Fe	Pyroxene	4
Ca	Calcite	3
Ca S Si	Gypsum	1
Ca Si Fe	(Tremolite/Actinolite)	2
Ti Si	(Rutile)	2
Ti Fe Si		0.5
K Cl Si		0.5
Pb Cl Br		3
Pb Si		3
Fe Zn Si S		1
S Si Na		1
Unassigned		4

[a]Si indicates that Si may be present in the particles or may be due to a spectral artifact (carbon absorption edge).
[b]( ) indicates only a possible mineral assignment for the cluster.

In a further test of the clustering procedure, analyses of particles of standard clay minerals, ripidolite, montmorillonite, nontronite as well as muscovite mica, were clustered. The procedure easily identified the different minerals, giving rise to well resolved clusters. These results, and results from other standard mineral particles, were compared to the clusters determined from the Phoenix aerosol and listed in Table I. This comparison indicated that, while many clusters were well resolved (e.g., those mentioned above), the alumino-silicate clusters in the Phoenix samples were probably mixtures of several mineral types. The minerals indicated in Table I have been identified in the Phoenix aerosol in the 5 to 50 um diameter size range (13). They were listed not as absolute assignments but as suggestions for the most prominent mineral type in the given cluster. Obviously, many of the particles were not necessarily crustal in origin. For example, there are many sources of iron and iron oxide particles other than magnetite. Also, evidence from other sites indicated that the titanium cluster may result from an anthropogenic source.

Table II. Classification Results for Chandler, Arizona, as percent of total particles classified.

Date	Quartz	Orthoclase	Muscovite	Calcite	Pyroxenes
Feb 22	8.8	8.3	21.3	1.3	4.0
23	10.0	7.5	22.0	4.3	5.8
24	8.1	8.7	32.0	2.2	4.2
26	11.9	6.5	24.7	2.0	1.7
27	15.8	7.7	18.2	1.1	1.7
28	10.6	9.4	21.6	1.4	2.0
Mar 3	10.6	5.6	19.3	8.0	5.2
4	10.6	6.4	22.5	3.5	1.8

Using the particle types outlined in Table I, a series of samples from Chandler, Arizona, were classified. The samples were collected over a two- week period in late February and early March. The results for several particle types are listed in Table II. The first interesting result is that muscovite is always more abundant then quartz, in contrast with the downtown Phoenix sample. In addition, the pyroxene, muscovite and calcite types are negatively correlated, over time, with quartz. The classification results were used as input for principal components analysis, with the observations being the different samples and the variables the particle types. The first principal component has a predominant weighting on muscovite, explaining 52% of the variance of the data set. The second principal component has strong positive weightings on the pyroxenes and calcite and strong negative weightings on quartz, explaining 23% of the variance. Therefore the crustal particles show a striking difference in behavior, counter to what one would have expected. This does not appear to be simply random behavior because the sample scores on principal component two show a good correlation with the east-west direction of upper level winds.

Summary

K-means cluster analysis is an excellent method for the reduction of individual-particle data, if extra clusters are used to allow for the non-spherical shape and natural variability of atmospheric particles. The "merge" method for choosing seedpoints is useful for detecting the types of low abundance particles that are interesting for urban atmospheric studies. Application to the Phoenix aerosol suggests that the ability to discriminate between various types of crustal particles may yield valuable information in addition to that derived from particle types more commonly associated with anthropogenic activity.

Acknowledgments

Financial support for this work was provided by grants ATM-8022849 and ATM-8404022 from the Atmospheric Chemistry Division of the National Science Foundation.

Literature Cited

1. Post, J. T.; Buseck, P. R. Environ. Sci. Technol., 1984, 18, 35-42.
2. Armstrong, J. T., Buseck, P. R. Electron Microsc. X-Ray Appl. Environ. Occup. Health Anal., [Symp.], [2nd], 1978, 211-228.
3. Bradley, J. P., Goodman, P., Chan, I. Y. T., Buseck, P. R. Environ. Sci. Technol., 1981, 15, 1208-1212.
4. Bradley, J. P., Buseck, P. R. Nature, 1983, 306, 770-772.
5. Buseck, P. R. and Bradley, J. P. In "Heterogeneous Atmospheric Chemistry"; Schryer, D. R., Ed.; GEOPHYS. MONOGR. No. 26, Am. Geophys. Union: Washington, D.C., 1982; pp. 57-76.
6. Thomas, E. and Buseck, P. R., Atmospheric Environment, 1983, 17, 2299-2301.
7. Anderberg, M. R. "Cluster Analysis for Application"; Academic Press: New York, 1973.
8. Massart, D. L.; Kaufman, L. "The Interpretation of Analytical Chemical Data by the Use of Cluster Analysis"; Wiley: New York, 1983; p. 107.
9. SAS Institute Inc. "SAS User's Guide: Statistics"; SAS Institute Inc: Cary, NC, 1982; pp. 417-434.
10. Tou, J. T.; Gonzalez, R. C. "Pattern Recognition Principles"; Addison-Wesley: Reading, MA, 1974; pp. 90-92.
11. Hartigan, J. A. "Clustering Algorithms"; Wiley: New York, 1975, p. 97.
12. Engelman, L.; Hartigan, J. A. J. Am. Stat. Assoc. 1969, 64, 1647-1648.
13. Pewe, T. L.; Pewe, E. A.; Pewe, R. H.; Journaux, A.; Slatt, R. M. Spec. Pap.--Geol. Soc. Am. 1981, No. 186.
14. Friedman, H. P.; Rubin, J. J. Am. Stat. Assoc. 1967, 62, 1159-1178.

RECEIVED July 17, 1985

Monitoring Polycyclic Aromatic Hydrocarbons

An Environmental Application of Fuzzy C-Varieties Pattern Recognition

R. W. Gunderson[1] and K. Thrane[2]

[1] Department of Mathematics, Utah State University, Logan, UT 84322
[2] Norsk Institutt for Luftforskning, N-2001 Lillestrøm, Norway

Data collected in a two year program to monitor polycyclic aromatic hydrocarbons in the vicinity of Sundsvall, Sweden, were analyzed by the Norwegian Institute for Air Research (NILU) to (1) determine the possibility of identifying emission sources, and (2) quantify the contribution from a local aluminum plant. NILU used the fuzzy c-varieties clustering program FOSE as one of two methods for carrying out the investigation. The original study was repeated using recent improvements in the fuzzy c-varieties technique and using the recursive clustering option and a quantitative measure of cluster quality which are features of a new program, FCVPC. The results of of these two investigations are discussed and compared. In general, the results and conclusions reached were in good agreement.

In 1978, the emission of benzo(a)pyrene (BaP) from an aluminum plant in the vicinity of Sundsvall, Sweden, was estimated to be about four times the total amount emitted from all the motor vehicles in that country. As might be expected, the result of this estimate caused considerable concern, and a survey of the air quality in the Sundsvall area was made in 1980-81. The program monitored concentrations of polycyclic aromatic hydrocarbons (PAH) and fluoride in ambient air, with samples being collected once each week. Concentrations of fluoride and meteorological data were measured by the aluminum company laboratory, while PAH concentrations were determined by the Norwegian Institute for Air Research (NILU).

0097–6156/85/0292–0130$06.00/0

In addition to studying the behaviour and transport of the PAH compounds, NILU was asked to study the possibilities of identifying the main emission sources of PAH, and of quantifying the contributions from the aluminum plant. NILU employed two methods of investigating these questions. One of those approaches involved the application of a relatively new family of non-hierarchical clustering algorithms, the fuzzy c-varieties (FCV) algorithms (Bezdek, et.al., 1).

One purpose of this paper is to describe the FCV clustering algorithms, using the NILU investigation as an example of their application to environmental chemometrics. Another purpose is to introduce a new measure of cluster validity for these algorithms. This measure, the validity discriminant, provides a quantitative replacement for the subjective evaluations of cluster quality which have been required in previous applications of FCV clustering, including the original NILU study. A part of the NILU study has been re-investigated using the validity discriminant, both to illustrate its use and to attempt to further validate the conclusions of that investigation.

The FCV Clustering Method

In the following, let $X = \{x_1,..,x_n\}$ denote a data set consisting of n measurement vectors, x_k, each with d features (attributes), x_{ki}; i.e., $x_k = (x_{k1},..,x_{kd})$.

There is a basic difference between most of the better known methods of cluster analysis and the FCV family of clustering algorithms. That difference concerns the traditional requirement that every measurement vector be eventually assigned to one, and only one, of the cluster classes. In FCV clustering, that requirement is replaced with a pair of conditions:

(1) that a measurement vector may simultaneously "belong" to more than one of the data classes, with its degree of membership in a particular cluster being represented by some value in the interval $[0,1]$;

(2) that the total "membership" of a given measurement vector over all the clusters must sum to unity.

If the notation $u_{ik} = u_i(x_k)$ is used to represent the degree of membership of the measurement vector x_k (k=1,2,..,n) in cluster i (i=1,2,..,c), then the two conditions above can be given a convenient mathematical statement:

(1') $0 \leq u_{ik} \leq 1$ (for every i and k)

(2') $\displaystyle\sum_{i=1}^{c} u_{ik} = 1$,   (for every k)

where c is the prespecified number of cluster classes.

The "fuzzy" designation for these algorithms is derived from the concept of shared membership. Ruspini (2) provided a pioneering application of fuzzy clustering, when he recognized that the formal fuzzy set logic introduced by Zadeh (3) seemed ideally suited for the imprecision encountered in cluster models. It was not until Dunn and Bezdek (4) published their Fuzzy ISODATA algorithm, however, before the applicability and usefulness of this type of clustering procedure became more widely appreciated. The FCV algorithms were first published in the paper by Bezdek et.al. (1), generalizing the Fuzzy ISODATA algorithms.

Since that time they have been used in a wide variety of cluster applications, and have provided a very powerful new technique for multivariable data analysis.

An r-dimensional linear variety in feature space $\mathbb{R}_d$ can formally be defined by an equation of the form

(3) $V(v;d_1,d_2,..,d_r) = \{ y \varepsilon \mathbb{R}_d \mid y = v + \displaystyle\sum_{j=1}^{r} t_j d_j \}$

where $0 < r < d$; v is a fixed vector in $\mathbb{R}_d$; the scalars t_j independently take on all values in $(-\infty, +\infty)$; and the spanning vectors d_j form a set of r orthonormal vectors in $\mathbb{R}_d$. In the case that $r=0$, the linear variety consists only of the point v in $\mathbb{R}_d$. If $r=1$, the linear variety forms a line through v in the direction d_1. If $r=2$, the linear variety consists of all of the vectors falling in the plane containing v and defined by d_1 and d_2; and so on. The euclidean orthogonal distance of a measurement vector x_k from a linear variety V_i is defined by the equation

(4) $D_{ik} = D(x_k,V_i) = \left((x_k-v_i)^T (x_k-v_i) - \displaystyle\sum_{j=1}^{r} d_j^T (x_k-v_i)(x_k-v_i)^T d_j \right)^{1/2}$

The geometric interpretation of the Euclidean orthogonal distance of a point to a line is shown in Figure 1.

With these definitions, the mathematical derivation of the FCV family of clustering algorithms depends upon minimizing the generalized weighted sum-of-squared-error objective functional

$$(5) \qquad J(U,\overline{V}) = \sum_{i=1}^{c} \sum_{k=1}^{n} (u_{ik})^m (D_{ik})^2$$

where $\overline{V} = \{V_1, V_2, .., V_c\}$ is a set of c linear varieties, all of the same dimension, and the minimization is carried out for a fixed $1 \leq m < \infty$ over all membership values u_{ik} and families of c linear varieties, $\overline{V}$. The value for m is usually chosen as m=2. Higher values may be chosen if it is desired to attach less weight to the importance of small membership values. If m=1, the solution of the minimization problem reduces to the special case where all of the u_{ik} are either 1 or 0 (i.e., a "hard", as opposed to "fuzzy", solution; Bezdek, et.al., 1).

Necessary conditions for minimizing Equation 5 are given by the equations (Bezdek, et.al., 1):

$$(6) \qquad u_{ik} = 1 \left/ \sum_{j=1}^{c} (D_{ik}/D_{jk})^{1/(m-1)} \right. , \quad \forall \ (i,k)$$

$$(7) \qquad v_i = \sum_{k=1}^{n} (u_{ik})^m x_k \left/ \sum_{k=1}^{n} (u_{ik})^m \right. , \ \forall \ i$$

where the spanning vectors $(d_{i1}, d_{i2}, .., d_{ir})$ required for computing the distances in Equation 6 are the r unit eigenvectors corresponding to the r largest eigenvalues of the, c, "fuzzy" scatter matrices

$$(8) \qquad W_i = \sum_{k=1}^{n} (u_{ik})^m (x_k - v_i)(x_k - v_i)^T \quad (i = 1, 2, .., c).$$

The FCV family of algorithms consist of a Picard iteration to an approximate solution of these equations, with the singular case (where at least one of the distances $D_{ik} = 0$) taken care of separately.

In order to get an idea of how the algorithms actually work, let us suppose it is desired to find c=2 linear clusters (r=1) for the data of Figure 2. The iterative solution of Equations 6-8 is initiated by guessing a starting membership matrix

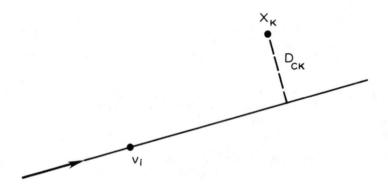

Figure 1. Orthagonal distance D_{CK} from measurement vector X_K to line through class weighted center v_i.

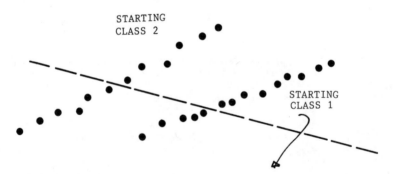

Figure 2. Two linear clusters, showing starting class membership assignments.

$$(9) \quad U_0 = \begin{bmatrix} u_{11} & u_{12} & \cdot\cdot & u_{1k} & \cdot\cdot & u_{1n} \\ u_{21} & u_{22} & \cdot\cdot & u_{2k} & \cdot\cdot & u_{2n} \end{bmatrix}$$

As an example, suppose the starting guess is to partition the data of Figure 2 into the two "hard" clusters indicated by the dashed lines (i.e., there is no "shared" membership). The matrix U_0 , in this case, would consist of all 1's or 0's.

The first step is to use Equation 7 to compute the two weighted means v_1 and v_2. Since all of the starting u_{ik} are either 0 or 1, in this case v_1 and v_2 are just the usual mean values, or centers, of the corresponding hard partition of the input data. When the u_{ik} include values between 0 and 1, the v_i's are often referred to as "fuzzy" cluster centers.

Having computed the centers v_1 and v_2, the next step is to find a best fit of the data in each cluster to lines running through the respective centers. This is done by computing the weighted scatter matrices of Equation 8. The eigenvectors of those matrices define the directions of the lines. A connection with the ideas of principal component analysis may be noted at this point. The idea is pursued further by Gunderson and Jacobsen (5).

The final step in the first iteration is to compute new membership values u_{ik} using Equation 6 and the distance definition D_{ik} of Equation 4.

The algorithm continues by using the new membership matrix as the starting membership matrix for a second iteration through the same equations. The iterations are allowed to continue until a stopping condition is reached; usually when the maximum change in membership values from one iteration to the next is less than some prespecified threshold value. Windham (6) showed that the iterations will always reach such a stopping point, regardless of how small the threshold is set.

The algorithms can be interpreted in this case as trying to obtain a "best" simultaneous fit of the data to two straight lines (in the sense of minimizing Equation 5). If r had been chosen as $r=0$ in the above problem, the distances D_{ik} would have reduced to just the usual Euclidean distance of the measurement vectors to the respective centers, and the resulting cluster "shapes" would have been forced to a best "round",

instead of linear, fit of the data. In general, it is
possible using the FCV algorithms to look for a best
simultaneous fit of the data to any c, linear,
hypersurfaces of common dimension $r < d$, by setting r to
an appropriate value at the beginning of the clustering
run. This a very valuable feature, in that it allows the
investigator greater flexibility in attempting to seek
out the structure in the data than is available with
most other clustering methods. An adaptive version of
the FCV algorithms was provided by Gunderson (7), which
is capable of seeking out clusters of mixed linear
shapes in the same data set.

The computation of the shared membership values u_{ik}
by Equation 6 usually results in relatively small values
being assigned to outliers or noisy measurement vectors.
It is not difficult to locate these values in the final
membership matrix, and the corresponding data vectors
can be singled out for deletion, or closer examination.

The u_{ik} values played an important role in the NILU
investigation. They were used to assess the relative
"quality" of a cluster configuration relative to
competing configurations. By clustering the data to
different configurations, and comparing the amount of
sharing between clusters in each case, a subjective
opinion could be formed as to which seemed to provide
the most natural partitioning. In the next section we
describe a more quantitative approach directed toward
answering the same question, which also makes use of
shared membership weighting.

The Validity Discriminant

The validity discriminant discussed in this section is
the descendant of an earlier cluster validity measure
used by Gunderson (8) to assess the quality of cluster
configurations obtained in an application of the Fuzzy
ISODATA algorithms. It is closely related to a method
suggested by Sneath (9) for testing the distinctness,
i.e. separation, of two clusters, and also borrows from
the ideas of Fisher's linear discriminant theory (see
chapt. 4, Duda and Hart,(10). The validity discriminant
attempts to measure the separation between the classes
of a cluster configuration usually, but not necessarily,
obtained by application of the FCV algorithms. A brief
description follows:

It is assumed that the membership values for all of
the measurement vectors, and the weighted centers, v_i,
are known for an arbitrary pair (i,j) of the c classes.
In analogy with Fisher's linear discriminant theory,
define the weighted (fuzzy) total scatter of the pair
(i,j) by

$$(10) \quad T_{ij} = \sum_{k=1}^{n} \left[(u_{ik})^m + (u_{jk})^m \right] (x_k - m_{ij})(x_k - m_{ij})^T \; ,$$

where the "pair center", m_{ij}, is given by

$$(11) \quad m_{ij} = \sum_{k=1}^{n} \left[(u_{ik})^m + (u_{jk})^m \right] x_k \; \div$$

$$\sum_{k=1}^{n} \left[(u_{ik})^m + (u_{jk})^m \right]$$

The weighted (fuzzy) within-cluster scatter of the pair (i,j) is defined by

$$(12) \quad W_{ij} = \sum_{k=1}^{n} (u_{ik})^m (x_k - v_i)(x_k - v_i)^T$$

$$+ \sum_{k=1}^{n} (u_{jk})^m (x_k - v_j)(x_k - v_j)^T$$

and the weighted (fuzzy) between-cluster scatter of the pair by

$$(13) \quad B_{ij} = \left[\left(\sum_{k=1}^{n} (u_{ik})^m \; \sum_{k=1}^{n} (u_{jk})^m \right) \div \right.$$

$$\left. (\sum_{k=1}^{n} (u_{ik})^m + \sum_{k=1}^{n} (u_{jk})^m) \right] (v_i - v_j)^T$$

It is not difficult to show that

$$(14) \quad T_{ij} = W_{ij} + B_{ij}$$

The validity discriminant is then defined by the extremum problem

$$(15) \quad V_d = \underset{z \neq 0}{\text{maximize}} \; \{ (z^T B_{ij} z) \big/ (z^T W_{ij} z) \}$$

where z is any vector in measurement space $\mathbb{R}_d$. It is well known that the solution to this problem can be obtained by solving the generalized eigenvalue problem

$$(16) \quad B_{ij} z = \lambda W_{ij} z$$

The largest eigenvalue so obtained is the maximal value of the ratio, and thus provides the value of the validity discriminant V_d .

The denominator of the ratio can be interpreted as the weighted scatter of the two classes about their respective centers, as measured on a line in the direction of the maximizing eigenvector z. The numerator measures the distance between the two cluster centers, projected upon the same line.

The objective in solving the maximization problem is to find that line on which this separation ratio has its best, or highest, value. The separation between all pairs in a given clustering can then be compared to the separation between cluster pairs for a competing configuration.

Such a measure of the separation between classes will work best when it can be assumed that the classes approximate multivariate normal distributions. That is a reasonable assumption for the classes modeled by the output of the FCV algorithms.

As a final remark, note that the membership weighting values in the definition of the validity discriminant are also raised to the power m. Usually this value will be chosen to be the same as that used to accomplish the FCV clustering being evaluated. It has been suggested, however, that a further qualitative indication of cluster quality may be obtained by comparing the values of the cluster discriminant obtained by raising m to consecutively higher powers. If there is little change in the values, the conclusion is that "most" of the data have shared membership values close to either zero or unity, i.e. a configuration of relatively good quality. A marked increase in the values as m increases would be taken as an indication that the sharing between classes is substantial and that it has a noticeable effect on the class models. Such a configuration would be considered of poor quality and rejected.

Application of FCV Cluster Analysis to Monitoring Polycyclic Aromatic Hydrocarbons

NILU used the FCV clustering method as one of two methods to carry out their objectives of:

(1) Studying the possibility of identifying the main PAH emission sources in the Sundsvall, Sweden, area; and

(2) Attempting to quantify the PAH contributions from the aluminum plant.

Table I. PAH Compounds Selected for FCV Clustering

Variable description
Biphenyl
Acenaphthene
Fluorene
Phenanthrene
Anthracene
Fluoranthene
Pyrene
Benzo(a)anthracene
BeP
BaP
Coronene

All of the compounds measured in the monitoring program are listed in the report by Thrane (11). Table I lists the compounds which were selected as variables for the cluster analysis. Feature (i.e. attribute) selection for the cluster analysis was partially based upon the results of a principal component analysis (Henry, 12). Additional features were included if (1) the compound occurred in relatively large concentrations, or (2), if a compound was known to have adverse health effect. Wind direction, wind speed, and temperature were recorded as ordered variables. The chemical measurements were taken at five locations. Descriptions of those sites and of the methods and techniques used to collect the data are described in detail in the report by Thrane.

The methodology of the investigation was based on the assumptions that:

(1) The ratio of fluoride to total PAH measured at a site, and wind direction, could be used as good indicators for the presence of aluminum industry emissions;

(2) Aluminum industry PAH contributions at a given site could be estimated by the following procedure:

(a) Using FCV cluster analysis to identify clusters whose centers could be associated with aluminum plant emissions;

(b) Multiplying the membership value for each measurement vector in the clusters associated with aluminum plant emissions by the total PAH measurement of the vector;

(c) Dividing the sum of the weighted PAH estimates for the aluminum plant clusters by the total measured PAH contribution at the site.

Because the FCV algorithms are basically a non-statistical approach to cluster analysis it was not possible to attach estimates of misclassification error for step 2a). Similarly, the amount of data which could be collected for the investigation was not considered sufficient for use in defining confidence levels for the absolute values of plant emissions determined by step 2b). These qualifications need to be taken into consideration when interpreting the results of the investigation.

Contributions were estimated at four different stations. For the first half of the monitoring program, sampling at these sites was carried out at each of the sites over 12-hour (day, and night) periods. Because of the expense, and also because of difficulties experienced trying to obtain sufficient amounts of PAH for the analysis during the winter months, it was decided later in the monitoring program to change to 24-hour sampling periods. The 24-hour samples were collected at times when there was little land- or sea-breeze.

The original study by NILU included the data from all three of the different sampling time-periods. The raw input data for each run corresponding to a particular time-period at a particular one of the four sites consisted of less than forty measurement vectors, with a maximum of twelve features (if wind direction were included) each. Subjective evaluation of the quality of various configurations was based on the relative extent of membership sharing between cluster classes, as measured by the membership values u_{ik}, and led to five clusters being used for each site in the analyses described above. The computer program, FOSE, used by NILU was an early implementation of the FCV algorithms. It included only one option for scaling the input data; by normalizing the data set to reference concentrations. However,normalizing the input data seemed to have little effect upon the clustering results obtained in this application.

Table II summarizes the estimates obtained in the
NILU study when using the methodology outlined above.
Results for the Haga site have been boxed in for later
comparison with results obtained using the cluster
validity discriminant. The lower values for the 24h
samples probably reflects their being collected when
there was little land- or sea-breeze for transport of
the emissions from the smelter. Daytime sea breezes
would tend to transport emissions toward and past the
Haga site, while the evening landbreezes would tend to
transport emissions back toward the Haga site.

Table II. NILU Estimated Contributions From Aluminum
          Plant Emissions

(FCV Cluster Analysis Using Program FOSE)

Station	Sample type	%	Estimate based on membership in clusters #
Kubikenborg	Day	83	1,2,3,5
	Night	86	1,3,4,5
	24h	75	1,2,4,5
HAGA	Day	87	1,2,3,5
	Night	84	1,2,4,5
	24h	72	2,3,4,5
Kopmansgatan	Day	46	1,4,5
	Night	54	1,2,5
	24h	50	1,3,4
Sidsjon	Day	57	3,4,5
	Night	57	1,4,5
	24h	52	2,4,5

For comparison purposes, Table III shows the results of
a second, non-clustering, method which was used by NILU
to obtain contribution

Table III. Estimated BaP Contribution from Aluminum Plant at Four Stations (non-clustering)

Station	Contribution of BaP(ng/m^3)	% of measured BaP average concentration
Kubikenborg	6.4	123
Haga	4.0	98.8
Köpmansgatan	2.1	60.0
Sidsjön	1.38	68.25

estimates at the same sites. This approach was based upon the assumption that the retention time of fluoride in the atmosphere is the same as the lifetime of BaP, and that the aluminum smelter was the only source of fluoride in the Sundsvall vicinity. The contribution of BaP from the smelter at the different stations was then estimated from the ratio of the BaP and fluoride emissions and the concentrations of fluoride in the air. It should be noted that the paucity of emission measurements made these estimates somewhat uncertain. That uncertainty is evidenced by the estimated contribution at the Kubikinborg site, which is 23% higher than the measured concentrations. Reasons were advanced by Thrane (11) which may explain these over-estimates.

Contributions At The Haga Site: The Validity Discriminant

As mentioned above, the NILU investigation was carried out using the program FOSE, which did not include the cluster validity discriminant, or a convenient option for recursive subclustering. In this section, a comparison of results is made between those obtained by the original investigation using FOSE, and results obtained using a new FCV clustering program (see the Appendix of this report) which does include these options. The Haga station was selected for the comparison study, since both of the methods used in the original investigation estimated a very high PAH contribution at that site from the aluminum smelter. The daytime 12-hour sampling-period data was selected because of the observed correlation between wind direction and the presence of high amounts of PAH. Along the eastern coast of Sweden, the sea-breeze starts about 9-11 a.m. The wind direction at the aluminum plant is within the sector from south to east. Highest concentrations of pollutants were found to occur at all four stations when the wind was coming from that sector. The 12-hour daytime data at the Haga site consisted of twenty seven measurement vectors.

The first step of the analysis was to use the FCV algorithms to partition the data into c=2, c=3, and c=4 classes. The columns under the heading for "original data" in Table IV show the resulting validity discriminant values. High values in these columns indicate relatively good separation between the pairs of clusters listed in the column to the left. The c=4 result for the original data showed cluster 2 split into two parts, while clusters 1 and 3 remained essentially unchanged. Since the case provided little additional information, it was not included in the table. The levels of each of the eleven chemical components

Table IV. Validity Coefficient Results

Pair	Original Data		Subcluster of Cluster #2	
	2 clusters	3 clusters	2 clusters	3 clusters
1-2	9.75	94.99	3.41	24.26
1-3		37.21		9.93
2-3		49.39		4.48

selected as features for the measurement vectors are listed in the first column of Table V. The center of the original data is shown in the first column, with the levels defining the centers of the three main clusters of Table IV shown in the next three columns. Using the same assumptions as in the NILU study, the fluoride levels present in the centers of clusters 1 and 3, relative to cluster 2, were used to suggest the association of those clusters with aluminum plant emissions. This association was strengthed by observing the average wind direction to be from the SE quadrant (a listing of wind direction for each collected sample can be found in Thrane's report,11).

Table V. Weighted Centers of Class Principal Component
         Models

Variable	Original Data	Three Clusters		
		1	2	3
Biphenyl	10.815	14.294	11.343	7.423
Acenapth	45.711	181.628	20.602	91.758
Fluorene	63.278	287.211	25.76	129.037
Phenanth	255.859	328.586	77.579	556.791
Anthrace	19.446	101.213	4.577	47.091
Fluoranth	129.372	681.924	33.573	299.236
Pyrene	79.172	406.757	21.152	184.203
BaA	13.65	57.692	2.608	39.235
BeP	20.124	127.031	3.173	38.964
BaP	8.17	47.611	1.295	17.441
Coronene	2.554	7.224	1.65	3.896

		Three Subclusters of Cluster #2		
		1	2	3
Biphenyl		10.663	11.345	7.043
Acenapth		46.731	11.158	12.68
Fluorene		48.314	15.37	22.8
Phenanth		160.945	30.176	87.628
Anthrace		10.868	1.517	4.363
Fluoranth		78.305	11.721	37.383
Pyrene		45.509	9.852	20.999
BaA		6.56	0.975	2.513
BeP		6.936	1.736	2.543
BaP		2.875	0.967	0.824
Coronene		1.321	2.071	0.814

Because cluster 2 split into two parts when the
total data was clustered to c=4, and because inspection
of the final membership matrix for the case c=3
indicated that this cluster was not so well-defined as
the other two, it was more closely examined using the
recursive clustering option available in the FCV
computer program used for this study (see the Appendix
for a brief discussion of the programs FCVPC and FCVAX).
In this option, a threshold is chosen which determines
those measurement with "sufficient membership" in a
given cluster to be re-clustered into additional
subclasses. Recursive clustering thus provides a
hierarchicalflavor to this otherwise partitional method.
The results of subclustering the second cluster are
shown in the remaining columns of Tables IV and V.
Better separation was obtained by subclustering to c=3,

rather than c=2. This would imply that a definite improvement in cluster quality is achieved by going to the case c=3. Subclustering the original cluster 2 to c=4 provided only a refinement of the case c=3. Looking at the centers of the subclusters in Table V, it appears reasonable to associate the first class with the aluminum smelter. This class also showed good correlation with the expected SE wind direction. The center of subclass 2 does not seem to be associated with aluminum production, a result further reflected by the class separation measured by the validity discriminant coefficient for classes 1 and 2. Class 3 summed to the smallest total PAH contribution   and was also the least distinct. The separation coefficient for all three subclasses of the original cluster 2 suggests that its center shows more similarity to the class which has little contribution from the smelter.

Conclusions

The two investigations of the Sundsvall data which have been discussed in this paper were separated by time, distance, and methodology. The major difference was in the approach used to assess the quality of competing cluster configurations. In the case of the NILU study, it was necessary to use a subjective procedure, based on inspecting the final membership matrix. A membership matrix with most elements near 0 and 1 was considered preferable to one whose elements indicated a greater extent of membership sharing between clusters. The more recent investigation relied upon a quantitative measure of cluster quality, the cluster validity discriminant, which was presented in this report.

Use of the subjective procedure may not be wholly inappropriate for investigations such as this one, where the total number of measurement vectors is relatively small. Typically, the procedure required comparison of the membership of 25-35 vectors in 2-4 classes for the NILU data. For studies where the data consists of many more measurement vectors, and may require investigating the existence of a number of classes in the data, such an approach may become impractical and the accuracy of the results questionable. This should not be the case for the validity discriminant, where the increase in measurement vectors and data classes results only in an increase in computation time.

The basic agreement achieved using the two different approaches was considered an important result in that it tended to validate the validity discriminant concept, and provided increased confidence in its use on larger data sets for other investigations.

The only serious disagreement between the two
investigations was in the interpretation of the
subcluster 3. The same class was identified as cluster 2
in the report by Thrane (1982), and was recognized as
being one whose association was unclear. That report
chose to associate the cluster with the aluminum plant,
while this investigation used the validity discriminant
measure to argue that it should not. The resulting final
estimate of aluminum plant contribution at the Haga site
was therefore slightly lower, at 82 %, than the 87% of
the original investigation.

It should be mentioned that the Haga data was also
clustered to linear and planar cluster shapes, but
neither case provided partitionings of quality
comparable to the round clusters reported in this paper.
Both subjective comparison of the "extent" of membership
sharing, and the filtering technique for the more
quantitative validity discriminant measure were used to
assess quality. Finally, only the centers information
provided by the principal component model for each class
was used in the original NILU investigation. Additional
between-class and within-class information could have
been extracted from the information presented by the
principal directions for each class, as well as the
scatter in those directions provided by the eigenvalues
of the class weighted covariance matrices.

Appendix: Programs FCVPC and FCVAX

A program which implements the FCV algorithms is
available in compiled versions for the VAX 11/780 and
for IBM PC and PC compatibles, with or without the 8087
coprocessor. The Pascal source code can be made
available for compilation on other machines. The
programs are fairly general in that options can be
selected from a menu which permits (1) computation of
the principal component model for the original data;(2)
an FCV clustering of the data;(3) recursive FCV
clustering of the data;(4) weighted maximum likelihood
classification of new sample data vectors; (5) and
adaptive weighted maximum likelihood classification.
Utility programs are included for manipulating the input
data and creating test data sets. The programs also
permit comparison of FCV clustering results with three
standard hierarchial clustering methods. The first
author should be contacted for further information.

Literature Cited

1. Bezdek,J.C.; Coray, C.; Gunderson, R.W.; Watson,
    J.D.; SIAM J. Appl. Math., 1981, 40 (Parts I and
    II), pp.339-372.

2. Ruspini, E., Inf. Sci., 1970, 2, pp.319-350.

3. Zadeh, L.A., Inf. and Cont., 1965, 8, pp.338-353.

4. Bezdek, J.C.; Dunn, J.C., IEEE Trans. Comp., 1975, C-24,pp.835-838.

5. Gunderson, R.W.; Jacobsen, T., Proc. of Nordic Symp. on Appl. Stat., 1983, pp.37-63.

6. Windham, M.P.,"Optimal FCV Clustering Algorithms", Utah State University Mathematics, 1983.

7. Gunderson, R.W., Int. J. Man-Machine Studies, 1983,19, pp.97-104.

8. Gunderson, R.W., Proc. 7th Tri-ennial World IFAC Congress, 1978, pp.1319-1323.

9. Sneath,P.H.A., Math. Geol., 1977, 9, pp.123-143.

10. Duda,R.O.; Hart, P.E., "Pattern Classification and Scene Analysis"; Wiley Interscience: New York, 1973; Chap. 4.

11. Thrane, K.E., "Polycyclic Aromatic Hydrocarbons in Ambient Air", Norsk Institutt for Luftforskning, 1982.

12. Henry, R., "Principal Component Analysis of PAH Data for Sundsvall", Norsk Institutt for Luftforskning, 1982.

RECEIVED July 17, 1985

Applications of Molecular Connectivity Indexes and Multivariate Analysis in Environmental Chemistry

Gerald J. Niemi[1], Ronald R. Regal[2], and Gilman D. Veith[3]

[1] Department of Pharmacology, University of Minnesota—Duluth, Duluth, MN 55812
[2] Department of Mathematical Sciences, University of Minnesota—Duluth, Duluth, MN 55812
[3] Environmental Research Laboratory, U.S. Environmental Protection Agency, Duluth, MN 55804

We have developed a data matrix of 90 variables calculated from molecular connectivity indices for 19,972 chemicals in the Toxic Substance Control Act (TSCA) inventory of industrial chemicals. Principal component analysis of this matrix revealed eight principal components that explained > 93 % of the variation in these data. The first three principal components convey generalized information on chemical structure: size, degree of branching, and number of cycles. The other components contained more specific information on branching, bonding, cyclicness, valency, and combinations of these structural attributes. Here we explored the use of the connectivity indices and their calculated principal components for their potential in predicting biodegradation as measured by biochemical oxygen demand (BOD) and the octanol/water partition coefficient. This approach showed promise in the prediction of biodegradation, but was of limited use in the prediction of the partition coefficient. Because it is possible to calculate the connectivity indices at a nominal cost for nearly all chemicals, the approach will prove especially useful for the identification of chemicals with similar structures and for systematically exploring where data are lacking on biological endpoints for chemicals in TSCA.

More than 50,000 chemicals are currently listed in the Toxic Substance Control Act (TSCA) inventory, but physical-chemical properties are available for a relatively small percentage and biological endpoints for even less. The costs associated with thoroughly testing all chemicals are prohibitive, so models are needed to (1) predict the environmental effects of a new chemical, or (2) assess whether the chemical should be subject to a detailed testing regime (1). Although models are available to

0097-6156/85/0292-0148$06.00/0

assess the environmental effects for some groups of chemicals (2-6), new compounds often cannot be categorized into one group or another. If it is unclear when a model can be applied to a new chemical, then the power of the model is substantially reduced.

To overcome this weakness, we are developing a quantitative structure–activity strategy that is conceptually applicable to all chemicals. To be applicable, at least three criteria are necessary. First, we must be able to calculate the descriptors or independent variables directly from the chemical structure and, presumably, at a reasonable cost. Second, the ability to calculate the variables should be possible for any chemical. Finally, and most importantly, the variables must be related to a parameter of interest so that the variables can be used to predict or classify the activity or behavior of the chemical (1). One important area of research is the development of new variables or descriptors that quantitatively describe the structure of a chemical. The development of these indices has progressed into the mathematical areas of graph theory and topology and a large number of potentially valuable molecular descriptors have been described (7-9). Our objective is not concerned with the development of new descriptors, but alternatively to explore the potential applications of a group of descriptors known as molecular connectivity indices (10).

Molecular connectivity indices are desirable as potential explanatory variables because they can be calculated for a nominal cost (fractions of a second by computer) and they describe fundamental relationships about chemical structure. That is, they describe how non–hydrogen atoms of a molecule are "connected". Here we are most concerned with the statistical properties of molecular connectivity indices for a large set of chemicals in TSCA and the presentation of the results of multivariate analyses using these indices as explanatory variables to understand several properties important to environmental chemists. We will focus on two properties for which we have a relatively large data base: (1) biodegradation as measured by the percentage of theoretical 5–day biochemical oxygen demand (BOD)(11), and (2) n–octanol/water partition coefficient or hereafter termed log P (12).

Data Base

The U.S. EPA Environmental Research Laboratory–Duluth with the help of their cooperators has developed a data matrix of 90 variables calculated from molecular connectivity indices (10) for 19,972 of the chemicals in TSCA. Molecular connectivity indices consist of four primary types (paths or the edges between atoms, clusters or branches, path/clusters, and cycles or rings) that are calculated from 0th to 9th order depending on the number of connections between atoms. Path terms can include as many orders as there are edges between atoms in the molecule, the minimum order for a cluster or a cycle is three, and the minimum for a path/cluster is four. Therefore, using 0th to 9th order, the number of variables for one set of connectivity indices is 30 variables. In our data base, we included three sets of connectivity indices: simple indices in which the molecule is assumed to be a saturated hydrocarbon, bond-corrected indices that adjust for double and triple bonds between atoms, and valence-corrected indices that adjust for the heteroatoms in the molecule.

The data base for biodegradation information consisted of 5-day BOD tests that were screened from the literature using a systematic review procedure (13). A total of 340 chemicals was included in this analysis. In contrast, the data base for log P values is much larger because Leo and Weininger (14) have developed an additive model to estimate log P for chemicals. Their model also gives an estimate for the confidence in the estimated value of log P. Here we used a data base of 1700 chemicals that were determined by Leo and Weininger (14) to have a high confidence in the estimated value of log P.

Statistical Analysis

We focused on four primary objectives: (1) calculation of the statistical properties of the molecular connectivity indices, (2) evaluation of how the dimensionality (90 variables) of the molecular connectivity variables could be reduced (15-17), (3) prediction of high or low 5-day BOD using the molecular connectivity variables as discriminators (18), and (4) estimation of log P values using the molecular connectivity indices as explanatory variables (19-20).

Evaluation of the statistical properties is a fundamental part of any statistical analysis and here we concentrated on the distribution of each variable. To reduce the dimensionality of this data set we used principal component analysis (PCA) to explore the covariance structure of these data and to reduce the variables to a more manageable number (PA1 method with no rotation, 21).

Because a complete explanation of the procedure used to predict the relative degree of biodegradation of chemicals is prohibitive (see 22), here we focus on the overall results. Briefly, our multivariate statistical analysis included (1) calculating the factor scores derived from six principal components of a principal component analysis extracted from 45 of the molecular connectivity variables for the 19,972 chemicals of TSCA, (2) using these six principal components to identify ten clusters in the principal component space with the K-means clustering algorithm of the Biomedical Computer Program (23), and (3) calculating the discriminant functions that best separated chemicals with high BOD values (theoretical BOD > 17%) from those with low BOD values (theoretical BOD < 13 %, there were no chemicals with BOD values between 13 and 17 %) using the molecular connectivity indices as discriminators (discriminant function analysis, 23). The BOD values were divided into high and low groups because we were most interested in the identification of whether a chemical was degradable (high BOD) or persistent (low BOD).

In comparison, log P does not follow a dichotomous logic similar to the BOD values and, therefore, we treated log P as a continuous variable. For the predictions of log P we explored several multiple regression models ($\underline{21}$) and used either principal components or molecular connectivity indices as explanatory variables. With the principal components we explored the use of non-linear trends by including second and third order polynomials as explanatory variables. With the molecular connectivity indices we used two different variable modifications in the regression models: (1) logarithmic transformations, and (2) standardization by size whereby the logarithm of the number of atoms was subtracted from the logarithm of each variable. The regression analyses included using both forward and stepwise methods. We were primarily concerned with finding combinations of variables that best reduced the standard error of the estimate (square root of the mean square error) ($\underline{21}$).

Results

Statistical properties of connectivity indices. Because there are so many variables involved in this analysis, we only included the mean and standard deviation of the original and log-transformed data for the simple molecular connectivity indices (Table I). The relative magnitude of the means and standard deviations for the bond-corrected and valence-corrected indices were very similar to those for the simple indices. There are two features of these variables that have an important bearing on their statistical distribution. First, each variable is skewed to the right because of the presence of a few large molecules relative to the bulk of the chemicals in the data base. Second, many of the variables have a relatively large number of zero values. This is especially pronounced in the third and fourth order cycle terms because more than 98 % of the industrial chemicals do not have this configuration. The log-transformations improve the distributions considerably in terms of the ratio between the mean and standard deviation, but many of the variables do not approximate a normal distribution. Given these problems in the distribution of the data, we proceed with caution and consider subsequent analyses as exploratory. Furthermore, the shortcomings in the distribution of these data supports the use of PCA in the data reduction as opposed to procedures of factor analysis. Because PCA is used in an exploratory manner and no specific hypothesis is being tested ($\underline{24}$), the assumption of normality in these data can be relaxed.

Data reduction. We used the log-transformed data in all analyses presented here. The PCA resulted in eight principal components with eigenvalues > 1 and they explained 93.5% of the variation in the original data (Table II). The first three principal components all convey generalized information on chemical structure: size (PC 1), degree of branchness (PC 2), and number of cycles (PC 3). PC 1 was positively correlated with all 90 variables ($\underline{r}$ > .32), except for the cyclic variables in which $\underline{r}$ was as low as .07 for the 3rd order cyclic variables. PC 2 was positively correlated ($\underline{r}$ > .26) with all cluster variables, but negatively correlated with all path and cyclic variables. PC 3

Table I. Statistical properties of 0th to 9th order for path,
cluster, path/cluster, and cycle types of simple connectivity
indices.

Term	Order	Original Mean	Original Standard deviation	Log-transformed[1] Mean	Log-transformed[1] Standard deviation	Percentage of zeroes
Path	0	11.24	5.35	1.09	.19	0.0
	1	7.10	3.51	.91	.20	0.0
	2	6.34	3.51	.86	.21	0.4
	3	4.67	3.00	.74	.23	4.9
	4	3.45	2.51	.63	.25	10.1
	5	2.51	2.15	.52	.26	19.0
	6	1.58	1.56	.39	.25	33.3
	7	1.00	1.19	.28	.23	38.5
	8	.61	.86	.20	.20	43.5
	9	.37	.63	.14	.17	50.1
Cluster	3	1.23	1.16	.32	.18	37.6
	4	.09	.21	.03	.07	80.2
	5	.36	.93	.11	.14	51.1
	6	.09	.47	.02	.09	82.8
	7	.18	.89	.04	.12	66.5
	8	.11	.83	.02	.10	89.5
	9	.13	.96	.02	.11	81.2
Path/ cluster	4	2.35	2.68	.47	.26	18.1
	5	3.23	3.86	.54	.32	18.2
	6	4.68	6.57	.61	.40	20.2
	7	5.43	8.74	.62	.45	22.8
	8	6.09	11.80	.60	.50	25.0
	9	6.50	15.09	.56	.54	29.3
Cycle	3	.01	.04	.00	.02	98.5
	4	.01	.05	.00	.02	98.1
	5	.02	.10	.01	.03	84.5
	6	.07	.19	.05	.06	36.2
	7	.12	.36	.08	.09	40.6
	8	.17	.59	.11	.13	45.9
	9	.21	.88	.13	.17	48.2

[1] natural logarithms

Table II. Interpretations of 8 principal components calculated from 90 variables based on molecular connectivity indices for 19,972 industrial chemicals.

Principal component	Eigen-value	Variation explained %	Low values of principal component	High values of principal component
1	47.36	52.6	small molecules	large molecules
2	12.14	13.5	few branches on molecule	multi-branched molecules
3	10.53	11.7	non-cyclic molecules	multi-cyclic molecules
4	5.19	5.8	7th to 9th order cycles	3rd to 4th order cycles
5	3.13	3.5	molecules with single bonds and simple branching patterns	multi-branched molecules with double or triple bonds and/or with many heteroatoms
6	2.83	3.2	complex branching patterns and multi-cyclic molecules with few heteroatoms	complex 3rd and 4th order cyclic molecules
7	1.74	1.9	5th to 7th order cycles	complex valence-corrected branched chemicals with many heteroatoms
8	1.22	1.4	short chain molecules with complex branches	long chain molecules with few heteroatoms

was positively correlated ($\underline{r}$ > .32) with all cyclic variables and negatively correlated with most of the other variables. The remaining principal components explained patterns of variation among chemicals in terms of gradients of cyclicness, bonding, branching, valency (e.g., number and positions of heteroatoms such as halogens and oxygen), and combinations of these structural attributes.

The results of the principal component analysis describe the "intrinsic" dimensionality that is measured about chemical structure by the molecular connectivity indices. Eight principal components explain 93.5 % of the variation in the 90 original variables, but 23 principal components are required to explain 99.0 % of the variation in these data. The relatively high percentage of explained variation in the first three principal components (77.8 %) indicates that there are relatively high correlations among the variables derived from the molecular connectivity indices and it is obvious that size, branchness, and cyclicness are major structural properties of chemicals. In terms of data reduction, the principal component analysis has reduced the dimensionality from 90 variables to 8 new variables that still explained 93.5 % of the variation in the data set. Furthermore, if we can interpret the first three principal components as a size axis (PC 1), an axis of branchness (PC 2), and an axis of cyclicness (PC 3), then we can envision the higher order components as conveying both subtle and potentially-important structural information because they are statistically uncorrelated ($\underline{r}$ = 0.0) or "independent" of size, branchness, and cyclicness.

Prediction of BOD value. In the ten clusters identified by the K-means clustering procedure, two clusters were represented by chemicals with only low BOD values and one cluster with nearly all (18 of 19 or 95 %) high BOD values (Table III). Therefore, no discrimination was attempted within these clusters. In the remaining clusters there were 202 high BOD chemicals and 97 low BOD chemicals. Of these, approximately 75 % (152 of 202) were correctly classified into the high BOD class, while 73 % (71 of 97) were correctly classified into the low BOD class. Using both the clustering and discrimination analyses, 77 % (170 of 220) and 78 % (93 of 120) of the chemicals in the data base were correctly classified. Within each of the clusters, between 2 and 4 molecular connectivity indices were used in the final discriminant functions to separate the two classes of BOD. Within each cluster a different combination of variables were used as discriminators. Because of the exploratory nature of this analysis, we lowered the F-ratio inclusion level to 1.0. In several of the clusters, the F-ratios for variables included in the discriminant functions were subsequently small (e.g., < 4.0).

Predictions of log P with regression. As would be expected, the largest values of the explained variation (r squared) and the smallest standard error of estimates found with the regression models were those that included all 90 variables. These models included one with log-transformed variables and the other with both log-transformations and with each variable standardized by size (Table IV). In general, the standard errors of the estimates were large relative to the mean which indicates a relatively poor fit for the models tested here.

TABLE III. Summary of cluster analysis and discriminant function analysis of high (> 17 % of theoretical) and low (< 13 % of theoretical) biochemical oxygen demand (BOD) values for 340 chemicals.

	Cluster analysis		Discrimination analysis	
	Number of chemicals		% Correctly classified	
Cluster	High BOD	Low BOD	High BOD	Low BOD
1	0	6	–	–
2	0	16	–	–
3	5	13	60	85
4	13	9	85	78
5	18	1	–	–
6	35	18	60	78
7	10	12	80	83
8	60	15	83	53
9	29	14	72	64
10	50	16	76	75
Total	220	120	77	78

Table IV. Summary of multiple regression analyses for the prediction of log P from molecular connectivity indices (sample size = 1700).

Variable		Regression Method	Over-all F	R^2 Variables 10	R^2 Variables All	Standard Error of Estimate
Number	Modification					
90	logarithm	stepwise	34.8	.54	.62	1.26
90	logarithm	forward	30.2	.41	.62	1.26
90	logarithm+ size	stepwise	35.3	.54	.62	1.26
90	logarithm+ size	forward	30.1	.27	.62	1.26
24	polynomials- third degree	stepwise	32.1	.24	.31	1.66
8	principal component	stepwise	64.0		.23	1.75

Discussion

Of the two exploratory analyses that we presented, one showed relatively promising results for the general classification of high or low BOD values while the other with log P provided unsatisfactory results. Among the advantages of the statistical procedures we used in the BOD model may have been the effect of reducing the dimensionality (PCA) and the complexity of the problem by clustering the chemicals into smaller groups. In contrast, for the regression analyses with log P, we attempted to deal with the problem on a global scale whereby all chemicals were included in one large analysis. Suckling et al. (25) have noted this dilemma by pointing out that complex problems are difficult to solve in one piece and need to be broken down into components, especially for the purposes of modelling. In the analytical procedures of the BOD model we used this process: (1) reduced the dimensionality of the data, (2) clustered the chemicals according to similar structural properties, and (3) attempted to analyze the parameter of interest within a specific cluster or group of similar chemicals. A further advantage of this procedure is that chemicals can be assigned independently to a cluster without prior biases regarding where it "fits" within the realm of existing chemicals. This is especially important from the perspective of environmental regulation of a new chemical. A newly designed chemical by virtue of being new does not necessarily fit into one group of chemicals or another.

Principal components. One of the potential powers of the principal components is that they are calculated from a large data base representing many chemical configurations and they are independent axes (intercorrelation r = 0.0)(24). Therefore, they are well-suited for building prediction models with multivariate methods such as regression where a desirable property among the explanatory variables is minimal multicollinearity (26). However, in these analyses, principal components and polynomial regression of the principal components were relatively poor predictors of log P and neither appear to be useful in this application. This is a contrast with the results presented by Murray et al. (19) who found a high correlation between log P and a form of the valence-corrected molecular connectivity index. Murray et al. (19) limited their correlation analysis to distinct chemical groups such as esters and alcohols. Our analysis included all the chemicals available in a global model.

Although the inclusion of all 90 variables produced a relatively high r^2 and the lowest standard error of estimate, we suspect that there are many spurious correlations involved in the prediction equation (27). It is slightly more encouraging that the first ten variables included in the stepwise regression models of the 90 variable log-transformed and 90 variable log-and size-transformed data sets produced r^2 values of .54. The standard error of the estimates of log P with the regression equations is substantially higher than the estimates obtained by Leo and Weininger (14) in their model.

Exploratory data analysis and data feedback. There are a large number of variables that can be calculated from a chemical structure, yet when the number of variables becomes large relative to the sample size then spurious or trivial correlations increase (27). For example, we believe the multiple regression analyses of log P with 90 variables may represent such a situation. Likewise, the DFA used in the BOD model examined a large number of variables relative to the sample size within some of the clusters (e.g., see cluster 3, Table III). In actuality, we used the DFA as an exploratory tool to find potential molecular configurations that were associated with high or low BOD values (22). This exploration led to the identification of several subgraphs of a molecule that may be associated with persistence or degradability of a chemical. The next step in the process would be to identify chemicals with these specific subgraphs and test them for their relative degradability.

We believe that two of the main limitations for progress in applications of multivariate analyses in environmental chemistry at the current time are (1) the lack of a large data base (hundreds or thousands of chemicals) that has been collected systematically for endpoints of environmental concern (e.g., biodegradation, toxicity, and carcinogenicity), and (2) the lack of a feedback mechanism whereby new chemicals or chemicals with specific molecular configurations are being retested for specific biological endpoints. Regarding the former, we recognize that there are many limitations in the BOD data base (e.g., 28); however, there are no alternative data bases with a sample size in the hundreds that can be used in attempts to model the important environmental process of biodegradation.

Future directions

We are pursuing analyses with a variety of additional endpoints, but it is already clear that connectivity indices and multivariate techniques will be useful in some applications and limited in others. A powerful practical application of this approach is that many variables can now be calculated for nearly all chemicals. Therefore, the "universe" of industrial chemicals in TSCA can be defined in multi-dimensional space. This space gives EPA a tool for the identification of (1) whether there is a similarly-structured chemical to a new chemical that needs to be evaluated, or (2) where data on biological endpoints are lacking within this chemical space. The former concept of similarly-structured or analogous chemicals having similar biological activity is at the heart of the structure-activity perspective for predicting the environmental effects of chemicals. The latter provides an effective, objective means for the identification of what chemicals need to be tested. Both tools will be useful in understanding the relationships between chemical structure and biological activity and in applications for the assessment of the environmental effects of chemicals. One of the most important aspects, however, will be to understand the capabilities of the approach and especially its limitations.

Kowalski (29) has correctly emphasized that "the measurements made by analytical chemists are associated with some degree of uncertainty," so "it is difficult to conceive of a more perfect marriage than analytical chemistry and statistics and applied mathematics." The marriage is the study of chemometrics and we are likely "only at the threshold of realizing the importance of statistical and mathematical techniques" (30) in applications of chemical measurements. Multitudes of data are currently capable of being generated by sophisticated analytical equipment. Similarly, adaptable software and high-speed computers are able to handle these large data sets. The next decade promises to be the "decade when chemistry advances as a multivariate science (31)."

Acknowledgments

This research was supported by a cooperative agreement (CR810824-0190) between the U.S. Environmental Protection Agency and the University of Minnesota, Duluth.

Literature Cited

1. Veith, G.D. "State-of-the-Art Report on Structure-activity Methods Development"; Environmental Research Laboratory, Duluth, 1980.
2. Ogino, A.; Matsumura, S.; Fujita, T. J. Med. Chem. 1980, 23, 437.
3. Yuan, M.; Jurs, P. Toxicol. and Appl. Pharmacol. 1980, 52, 294.
4. Ray, S. K.; Basak, S. C.; Raychaudhury, C.; Roy, A. B.; Ghosh, J.J. Ind. J. Chem. 1981, 20B, 894.
5. Birge, W. J.; Cassidy, R. A. Fund. and Appl. Toxicol. 1983, 3, 359.
6. Gillett, J. W. Environ. Toxicol. and Chem. 1983, 2, 463.
7. Balaban, A. T. Theoret. Chim. Acta. 1979, 53, 355.
8. Sabljic, A.; Trinajstic, N. Acta Pharm. Jugosl. 1981, 31, 189.
9. Bertz, S. H. "A Mathematical Model of Molecular Complexity"; King, R. B., Ed.; STUDIES IN PHYSICAL AND THEORETICAL CHEMISTRY Vol. 28, Elsevier Science Publishers, Amsterdam, 1983; pp. 206-221.
10. Kier, L. B.; Hall, L. H. "Molecular Connectivity in Chemistry and Drug Research"; Academic Press, New York, 1976, p. 257.
11. Sawyer, C. N.; Bradney, L. Sew. Works J. 1946, 18, 1113.
12. Hansch, C.; Quinlan, J. E.; Lawrence, G. L. J. Org. Chem. 1968, 33, 347.
13. Vaischnav, D. "Biochemical oxygen demand data base"; Call, D.J.; Brooke, L.T.; Vaischnav, D.; AQUATIC POLLUTANT HAZARD ASSESSMENT AND DEVELOPMENT OF HAZARD PREDICTION TECHNOLOGY BY QUANTITATIVE STRUCTURE-ACTIVITY RELATIONSHIPS. University of Wisconsin, Superior research project report (CR809234) 1984.
14. Leo, A.; Weininger, D. "CLOGP Version 3.2 User Reference

Manual"; Medicinal Chemistry Project, Pomona College, Claremont, 1984.
15. Cramer, R. D. J. Am. Chem. Soc. 1980, 102, 1837.
16. Lukovits, I.; Lopata, A. J. Med. Chem. 1980, 23, 449.
17. Burkhard, L. P.; Andren, A. W.; Armstrong, D. E. Chemosphere 1983, 12, 935.
18. Geating, J. "Project Summary: Literature Study of the Biodegradability of Chemicals in Water: Vols. 1 and 2"; U.S. Environmental Protection Agency, Cincinnati, EPA-600/s2-81-175/176, 1981, p. 4.
19. Murray, W. J.; Hall, L. H.; Kier, L. B. J. Pharm. Sci. 1979, 64, 1978.
20. Govers, H.; Ruepert, C.; Aiking, H. Chemosphere 1984, 13, 227.
21. Nie, N. H.; Hull, C. H.; Jenkins, J. G.; Steinbrenner, K.; Bent, D. H. "Statistical Package for the Social Sciences"; McGraw-Hill, New York, 1975, p. 675.
22. Niemi, G. J.; Regal, R. "Predicting Biodegradability from Chemical Structure: the Use of Multivariate Analysis"; U.S. Environmental Protection Agency, Duluth, 1983.
23. Dixon, W. J.; Brown, M. B. "Biomedical Computer Programs – P Series"; University of California Press, Berkeley, 1979, p. 880.
24. Morrison, D. F. "Multivariate Statistical Methods"; McGraw-Hill, New York, 1967, p. 338.
25. Suckling, C. J.; Suckling, K. E.; Suckling, C. W. "Chemistry Through Models"; Cambridge University Press, Cambridge, 1978.
26. Tatsuoka, M. "Multivariate Analysis"; J. Wiley, New York, 1971.
27. Wold, S.; Dunn, W. J. III. J. Chem. Inf. Comput. Sci. 1983, 23, 6.
28. Howard, P. H.; Saxena, J.; Sikka, H. Environ. Sci. Technol. 1978, 12, 398.
29. Kowalski, B. R. Anal. Chem. 1980, 52, 112R.
30. Kowalski, B. R. Anal. Chem. 1978, 50, 1309A.
31. Kowalski, B. R.; Wold, S. In "Handbook of Statistics, Vol. 2"; Krishnaiah, P. R.; Kanal, L. N. Eds.; North-Holland Publishing Co., 1982; Chap. 31.

RECEIVED June 28, 1985

Pattern Recognition of Fourier Transform IR Spectra of Organic Compounds

Donald S. Frankel

Center for Chemical and Environmental Physics, Aerodyne Research, Inc.,
Billerica, MA 01821

Two pattern recognition techniques are applied to the
analysis of the library of FTIR spectra compiled by the
US EPA. The patterns which emerge demonstrate the
influence of molecular structure on the spectra in a
way familiar to chemical spectroscopists. They are
also useful in evaluation of the library, which is not
error free, and in assessing the difficulties to be
expected when using FTIR spectra for complex mixture
analysis.

For testing the applicability of numerical pattern recognition to
realistic spectra, we chose to examine the increasingly relevant
problem of complex mixtures occurring in the polluted environment.
Several case studies have shown that chemical waste dump soil and
effluents, airborne particulates, municipal wastewater, river
sediments, etc. can often contain highly complex mixtures of organic
compounds.(1-5) Particles from combustion sources are also expected
to contain a broad variety of adsorbed species resulting from
incomplete burning of the fuel, which is often itself a complex
mixture.(6) Many excellent techniques have been developed for these
problems, including infrared spectroscopy, gas chromatography, mass
spectroscopy, high performance liquid chromatography, atomic
absorption spectroscopy etc.

However, problems with these techniques still exist. For
example, gas chromatography seldom produces complete component
separation when more than a few are present.(7) The problem of peak
overlap is widely recognized and not simple to deal with. The same
may be true of liquid chromatography. Traditional methods of
analysis of particles involve extraction of the organic material in a
variety of solvents before analysis. It is difficult to defend these
procedures if they are challenged (loss of material or contamination
during extraction, degradation of samples during storage, etc).(8)
If mass spectroscopy is the chosen method for the analysis of the
"resolved" components, one is faced with the difficult task of
interpreting the data.(9) All of these techniques are difficult,
time consuming and expensive.

0097-6156/85/0292-0160$06.00/0
© 1985 American Chemical Society

To improve this situation, we have sought a simple, nonintrusive
way to analyze mixtures. We chose Fourier Transform Infrared (FTIR)
Spectroscopy because the data are highly molecule specific and the
instruments are relatively simple to operate, disturb the sample very
little and obtain digitized data quickly. If the identities of the
components of a complex mixture are known, FTIR spectroscopy can be
used to analyze the mixture quantitatively. To analyze mixtures
whose components are unknown, we have adopted the alternative
approach of analyzing by chemical class. Infrared spectroscopists
have for several decades made use of the characteristic absorption
bands of functional groups to identify or classify unknowns.(10) In
this way, we hoped to identify and quantify the classes of organic
compounds in the spectrum of a mixture, rather than the specific
compounds. However, even this simplification would be virtually
impossible to implement were it not for the existence of high speed
digital computers and a set of statistical techniques known
collectively as pattern recognition. This paper describes the first
necessary steps toward our ultimate goal of mixture analysis by class
using pattern recognition techniques. We will examine groups of pure
compound spectra to see if they form distinct classes.

Pattern Recognition Background

Pattern recognition has received an increasing amount of well
deserved attention from chemists in recent years. Several excellent
review articles on chemical applications(11-13) and a number of
general texts have been published.(14-17)
      There are two necessary and related preconditions which must be
satisfied for complex mixture analysis by pattern recognition to be
successful. First, we must obtain an adequate data base of FTIR
spectra from which we can derive the spectral patterns we need to
recognize. Second, we must demonstrate that there is a suitable
measure or metric of similarity between the spectra. It is these two
conditions which were evaluated by the work presented here. Pattern
recognition techniques were most suitable for the evaluation.
      Pattern recognition techniques generally fall into two broad
categories, supervised and unsupervised. Unsupervised techniques
make no assumptions about the number of classes present in the data
set nor their relationship to one another. Rather, this information
is the hoped for result. In supervised learning, the number of
classes has already been determined, and a group of items whose class
assignment is known, called the "training set", is used to develop a
computational decision rule for assigning unknowns.
      The two pattern recognition techniques used in this work are
among those usually used for unsupervised learning. The results will
be examined for the clusters which arise from the analysis of the
data. On the other hand, the number of classes and a rule for
assigning compounds to each had already been determined by the
requirements of the mixture analysis problem. One might suppose that
a supervised approach would be more suitable. In our case, this is
not so because our aim is not to develop a classifier. Instead, we
wish to examine the data base of FTIR spectra and the metric to see
if they are adequate to help solve a more difficult problem, that of
analyzing complex mixtures by class.

Experimental

Classes, Data Base and Metric. The definition of a chemical class in
this work was predetermined by the type of result which was required
and was in keeping with that of other pattern recognition work. A
substance was assigned to a class based on its carbon skeleton and
functional groups. The hierarchy alkane < alkene < arene was
established such that a substance was assigned according to the
highest ranking skeletal unit it contained. Each distinct functional
group established a separate class, as did combinations of groups,
but multiple occurrence of the same group did not. Thus the alkyl
mono and polyalcohols were in one class, the mono and polychlorinated
alkanes were another class, and the alkanes with chloro and hydroxyl
substituents were in a third class.

Our library was compiled by the U.S. Environmental Protection
Agency and consists of 2300 FTIR spectra covering the 450-4000 cm^{-1}
range at 4 cm^{-1} resolution.(18) In the calculations presented here,
the resolution was reduced numerically to 16 cm^{-1}.

Each spectrum is regarded as a point in an N dimensional space.
The coordinates of each point are the absorbance values at each
wavenumber interval. Several metrics are widely used.(19) The first
is the N dimensional cartesian distance between points i and j,

$$d_{ij} = \left(\sum_k (x_{ik} - x_{jk})^2 \right)^{1/2} \tag{1}$$

where x_{ik} is the absorbance of compound i at wavenumber k. Since
closely related points produce a small distance, d_{ij} is often
converted to a similarity measure d^*_{ij} according to

$$d^*_{ij} = 1 - d_{ij}/d^{max}_{ij} \tag{2}$$

In the second metric, the one used in this work, attention is
focussed on the vector from the origin to each N dimensional
point.(20-22) The cosine of the angle between these vectors is given
by their dot product,

$$\cos \theta_{ij} = \sum_k x_{ik} \, x_{jk} \tag{3}$$

provided only that the vectors have been previously normalized to
unit length. This metric is already a similarity measure, in that
closely related spectra have this dot product near unity. Note also
that unlike peak position similarity measures, which are required by
some data bases,(23) this metric implicitly makes use of our
library's information about relative peak position, width and
intensity. Explicit lists of the compounds used in the calculations
described below may be obtained from the author on request.

Technique 1: Clustering of the Metric Matrix. The first step in
evaluating the library was to separate it into classes by explicitly
scanning the printed list of compounds. At this point it became

clear that some classes were represented by very few examples and had
to be dropped from further consideration. Unfortunately, among these
were the polychlorinated biphenyls (PCBs), and polycyclic aromatic
hydrocarbons (PAH).
    The further evaluation of the library and metric proceeds from
the metric matrix. Its elements are the dot products of each
spectrum with every other in its class.
    The first pattern recognition technique used was clustering of
the metric matrix, which requires no matrix inversion. Clustering
proceeds in the following way. The spectra producing the largest dot
product are linked and are treated thereafter as a unit by averaging
their matrix elements with the other spectra. The procedure is
repeated until all the spectra are linked.The results are displayed
graphically as a dendrogram.(24)

Technique 2: Eigenanalysis. It is well known that the structure of
a data set can be uncovered by performing an eigenanalysis of its
covariance matrix.(14) This is often called principal component
analysis. That is, we arrange the M measurement made on each of the
N objects as a column vector and combine them to form an M x N
matrix, A. A matrix B, resembling the covariance matrix of this
data set, is an M x M matrix AA^T whose elements are given by

$$B_{ij} = \sum_{k=1}^{N} A_{ik} A_{jk} = \sum_{k=1}^{N} A_{ik} A_{kj}^T \qquad (4)$$

    The eigenvectors of this matrix are linear combinations of the
measurements, and the eigenvalues are a direct measure of the
fraction of total variance accounted for by the corresponding
eigenvector. This analysis is the basis for the Karhunen–Loeve
transformation, in which the data are projected onto the plane of the
two eigenvectors with largest eigenvalue. This choice of axes
displays more of the data variance than any other.
    Closely related to this procedure is a less widely used analysis
based on the similarity matrix

$$S = A^T A \qquad (5)$$

which is of dimension N x N. The eigenvectors of this matrix yield
the linear combinations of objects which best represent the variance
of the entire data set.(26) Since each FTIR spectrum contains
hundreds or thousands of data points, using the similarity matrix
greatly reduces the size of the eigenvalue problem and gives results
equivalent to the more customary Karhunen–Loeve projection.

Results and Discussion

Dendrograms. In many cases, only one class was considered in a given
clustering calculation. Several types of information were obtained
in this way. First, it was verified that spectra of compounds in the
same class do indeed resemble each other in the dot product sense.
Second, blank or misclassified spectra could be easily recognized and
either discarded or reclassified, respectively. Thus, clustering

calculations can be a useful means of prescreening large libraries, which are often not error free. The clustering calculations also provide other kinds of information about chemical classification of infrared spectra. These include the effect of single substituents, carbon skeleton and multiple substitution. These effects were noticeable even under the more difficult circumstance wherein two classes were clustered simultaneously.

The clustering algorithm was applied to the spectra of the polycyclic aromatic alcohols and chlorides as shown in Table I. These calculations are encouraging for two reasons. First, they show that the first three spectra (alcohols) cluster closely with each other, as do the last three (chlorides). Second, the two groups cluster less well with each other than they do internally.

A similar clustering calculation was performed on the combined spectra of alkyl and aryl hydrocarbons. A schematic drawing of the dendrogram is shown in Figure 1. Eight branched, saturated hydrocarbons of C_5 to C_8 group together. Eight aromatics having one or more saturated side chains of three to five carbons form a second cluster. These two groups cluster with each other showing the influence of the C_3 - C_5 chains on the spectra. A group of 7 branched C_6-C_{12} aliphatics form a third cluster. Another group of aromatics having only one side chain or two short ones form a fourth cluster. These four clusters taken together are well separated from the remainder of the spectra, having a dot product with them of 0.11.

A group containing nearly 30 cyclic or long straight chain aliphatics cluster together. Generally speaking, this clustering reflects the decreasing relative abundance of methyl groups present in the molecules, going from C_5 - C_8 branched alkanes to long straight chain or cyclic alkanes.

The clustering of the remaining 20 spectra consists mainly of groups of two or three fairly similar spectra which join the other groups at much lower levels of similarity. The spectra in this region are largely benzene rings substituted with methyl groups, but include biphenyl, two of its derivatives and benzene itself. This result indicates that methylated benzenes, benzene and biphenyl are special cases. Whether this is the result of symmetry, the lack of aliphatic carbon–carbon single bonds or something else cannot be said at this time.

A similar calculation was performed for the aliphatic alcohols. Some subclustering could be seen in this case, but in general the aliphatic alcohols cluster at dot product levels around 0.90 or higher regardless of chain length and substitution pattern.

The next calculation attempts what has always been considered the most difficult problem that will be encountered in a mixture, namely, distinguishing singly and multiply substituted spectra. For this calculation, we have chosen the aromatic alcohols, chlorides and chlorinated alcohols.

The spectra separate into two large well separated groups. In the first, smaller clusters can be recognized: ortho diols, para–alkyl phenols, meta–alkyl phenols, ortho–chloro phenols, and multiply chlorinated benzenes. In the second group, these subclusters can be found: ortho–alkyl phenols, benzenes bearing chlorine on alkyl side chains, and ortho–chlorotoluenes.

Table 1 - Clustering of Chloro and Hydroxy Polycyclic Aromatics[a]

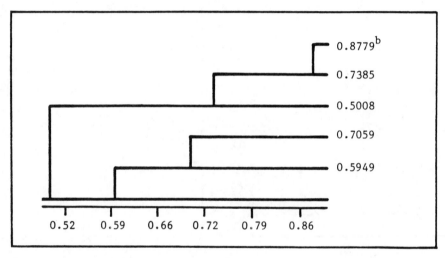

[a]The order of spectra is: 1,6 -naphthalenediol; 1-naphthol;
2-naphthol; 1-chloronaphthalene; 1-chloro, 2-methylnaphthalene;
2-chloronaphthalene.

[b]Numbers in this column reflect the degree to which the spectra or
groups of spectra are similar to the others. For example, the
number 0.8779 means that spectra 1 and 2 are similar at the level
0.8779. The number 0.7385 means that 1 and 2 together are similar
to 3 at a level of 0.7385. Numbers 1, 2, and 3 together are
similar to 4, 5, and 6 together at a level of 0.5008. The pattern
suggests two groups of three spectra.

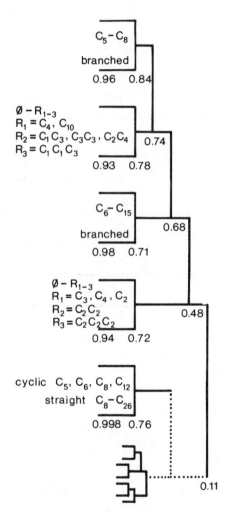

Figure 1.   Schematic Clustering of Alkyl and Aryl Hydrocarbons.
Numbers indicate the range of levels at which clustering occurs.
The group at the bottom contains benzene, biphenyl and their
methyl derivatives.

The clustering seems to indicate that ring substitution patterns
are recognized almost as easily as the substituents themselves. That
the clustering calculations display this information correlates with
the determination of aromatic ring substitution patterns from
infrared spectra practiced by chemists.(25) The results also
indicate that it may be best to consider the compounds with chlorines
on alkyl side chains as a separate class.

On the basis of these clustering results, the EPA library of
FTIR spectra was judged adequate as a source of spectra to form the
data base for the mixture analysis problem and the dot product was
deemed an adequate similarity measure. Every chemical class
considered to be a candidate for inclusion was subjected to the
clustering algorithm. Only those classes exhibiting a high degree of
internal similarity were retained in the mixture analysis data base.

Eigenvector Plots. Our prejudice was that distinguishing classes
with differing functional groups would be less difficult than
distinguishing different carbon skeletons. We therefore choose the
latter as a more stringent test of this means of data display.

The first example is for the alkyl and aryl chloro alcohols.
The plot using the first two eigenvectors is shown in Figure 2. As
is generally true of this analysis, the first eigenvector of the
similarity matrix is very nearly the average for all the objects and
is not very useful for separating the classes. In this case however,
the second axis is sufficient to show complete separation of the two
classes.

In the case of the alkyl and aryl hydrocarbons, where no
functional groups are present, the separation problem is more
difficult and the first and second eigenvectors do not yield
separation. However, it is necessary only to resort to the third
eigenvector to achieve nearly complete separation, as shown in
Figure 3.

The alkyl and aryl chloro spectra are an even more difficult
problem. The first two eigenvectors clearly separate one group of
alkylchloro compounds but leave another group and the arylchloro
compounds severely overlapped. Use of the third eigenvector, as
shown in Figure 4, leaves the separated group intact and better
separates the overlapped groups. Separation between these classes is
still far from complete and use of the fourth eigenvector does not
improve it. Examination of the plot shows that for these two
classes, the well separated alkyls all contain four carbon atoms or
more. Thus, an analysis of this type can be a guide for more
effective criteria of class membership.

The final example involves the alkenyl hydrocarbons, a class
whose distinguishing group absorptions are only subtly different from
those of other hydrocarbons. Figure 5 shows the projection of
alkenyl and aryl hydrocarbons onto the second and third
eigenvectors. Better separation occurs in this case than could be
obtained for the chloro compounds shown in Figure 4 using the third
and fourth eigenvectors. Thus, it may be expected that the alkenyl

Overall, the results show that the dot product metric, when used
with the US EPA data base of FTIR spectra, produces clusters of
compounds which make sense chemically. The data base and metric
should not therefore be impediments to the development of a pattern

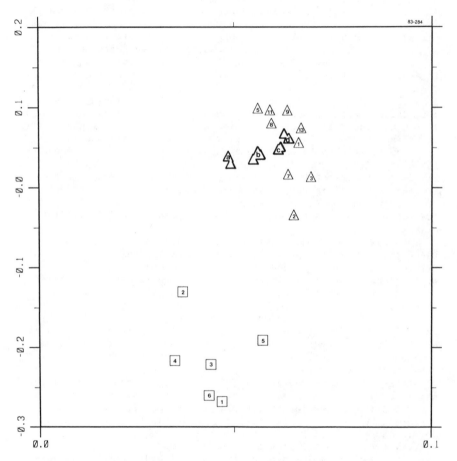

Figure 2. Alkyl Chloro Alcohols (□) and Aryl Chloro Alcohols
(Δ). Group a: Δ6, 15, 18; group b: Δ4, 14, 16; group c: Δ12,
17; group d: Δ10, 19.

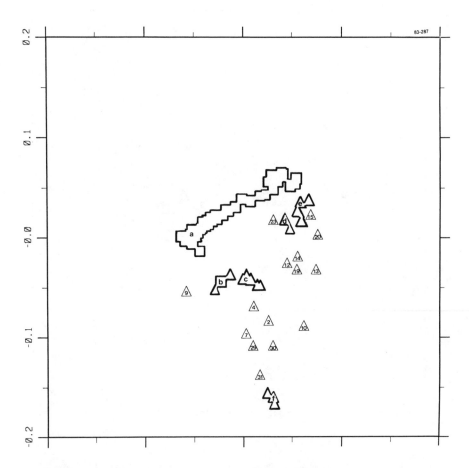

Figure 3.   Alkyl (□) and Aryl (Δ) Hydrocarbons.   Group a:□1-23,
25-48; group b:□24, Δ24, 25; group c: Δ1, 3, 8, 11, 18; group
d: Δ5, 6; group e: Δ17, 26, 27, 28; group f: Δ16, 21, 22.

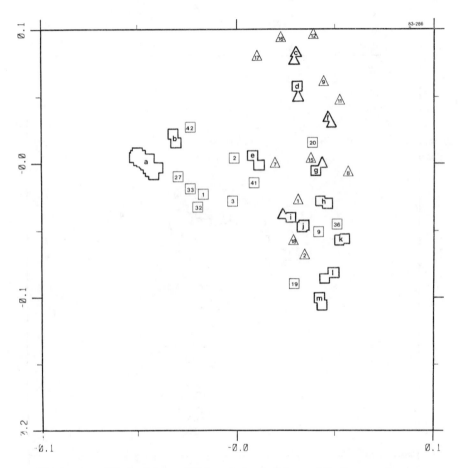

Figure 4.   Alkyl Chloro (□) and Aryl Chloro (Δ) Compounds.
Group a:□4, 11-14, 22-26, 30, 31, 34; group b: □6, 7;   group
c: Δ4, 6, 13; group d: Δ3, □10; group e: □5, 40;   group f:
Δ5, 14; group g: □8, Δ10; group h:□21, 37; group i: □39, Δ19;
group j: □15, 18; group k: □16, 28;   group 1: □35, 38; group
m:□17, 29.

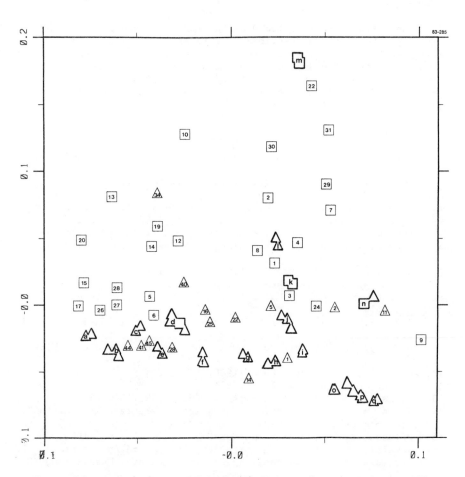

Figure 5.   Aryl (□) and Alkenyl (Δ) Hydrocarbons.   Group a: Δ29,
30; group b: Δ18-20; group c: Δ32, 46; group d:□ 23, Δ3, 4, 13;
group e: Δ27, 42; group f: Δ31, 38, 39; group g: Δ26, 37; group
h: Δ17, 36; group i: Δ8, 9; group j: Δ23, 24, 33; group k: □ 11,
18; group l: Δ6, 12; group m: □ 16, 21; group n: □ 25, Δ10;
group o: Δ13, 48; group p: Δ7, 15, 47, 49; group q: Δ21, 35.

recognition technique for the class analysis of complex mixtures. Such a technique has in fact been developed. A full description of it will appear in a future publication.

Acknowledgments

Work supported in part by the U.S. Environmental Protection Agency. "Although the research described in this article has been funded wholly or in part by the U.S. Environmental Protection Agency through Contract Numbers 68-02-3476 and 68-02-3751 to Aerodyne Research, Inc., it has not been subjected to Agency review and therefore does not necessarily reflect the views of the Agency and no official endorsement should be inferred."

Literature Cited

1.   Elder, V.A.; Proctor, B.L.; Hites, R.A. Envir. Sci. Tech. 1981, 15, p. 1237.
2.   Bopp, R.A.; Simpson, H.J.; Olsen, C.R.; Kostyk, N. Envir. Sci. Tech. 1981, 15, 211.
3.   Bombaugh, K.J.; Lee, K.W. Envir. Sci. Tech. 1981, 15, 1142.
4.   Gurka, D.F.; Betowksi, L.D. Anal. Chem. 1982, 54, 1819.
5.   Eganhouse, R.P.; Kaplan, I.R. Envir. Sci. Tech. 1982, 16, 541.
6.   Yergey, J.A.; Risby, T.H.; Lestz, S.S. Anal. Chem. 1982 54, 354.
7.   Davis, J.M.; Giddings, J.C. Anal. Chem. 1983, 55, 418; Rosenthal, D. Anal. Chem. 1982 54, 63.
8.   Ramdahl, R.; Becker, G.; Bjorseth, A. Envir. Sci. Tech. 1982, 16, 861.
9.   Jurs, P.C.; Kowalski, B.R.; Isenhour, T.L. Anal. Chem. 1969, 41, 21.
10. Bellamy, L. "The Infrared Spectra of Complex Molecules", Wiley, NY, 1954.
11. Kowalski, B.R.; Anal. Chem. 1980, 52, 112R.
12. Kryger, L. Talanta 1981, 28, 871.
13. Frank, I.E.; Kowalski, B.R. Anal. Chem. 1982, 54, 232R.
14. Fukunaga, K. "Introduction to Statistical Pattern Recognition", Academic Press: NY, 1972.
15. Nilsson, N.J. "Learning Machines", McGraw Hill: NY, 1965.
16. Kowalski, B.R. (ed.) "Chemometrics, Theory and Application", ACS Symposium Series 52, American Chemical Society: Washington, DC, 1977.
17. Jurs, P.C.; Isenhour, T.L. "Applications of Pattern Recognition", Wiley: NY, 1975.
18. Lowry, S.R.; Huppler, D.A. Anal. Chem. 1981, 53, 889.
19. Sogliero, G.; Eastwood, D.; Ehmer, R. Appl. Spectrosc. 1982, 36, 110.
20. Azzaraga, L.V.; Williams, R.R.; De Haseth, J.A. Appl. Spectrosc. 1981, 35, 466.
21. Kowalski, B.R.; Jurs, P.C.; Isenhour, T.L.; Reilley, C.N. Anal. Chem. 1969, 41, 1945.
22. Killeen, T.J.; Eastwood, D.; Hendrick, M.S. Talanta 1981, 28, 1.
23. Lowry, S.R.; Woodruff, H.B.; Ritter, G.L.; Isenhour, T.L. Anal. Chem. 1975, 47, 1126.

24. Davis, J.C., "Statistics and Data Analysis in Geology", Wiley, NY, 1973.
25. Roberts, J.D.; Caserio, M.C., "Basic Principles of Organic Chemistry,", Benjamin: NY, 1965.
26. Warner, I.M.; Christian, G.D.; Davidson, E.R.; Callis, J.B. Anal. Chem. 1977, 49, 564.

RECEIVED July 17, 1985

Use of Composited Samples To Increase the Precision and Probability of Detection of Toxic Chemicals

Gregory A. Mack[1] and Philip E. Robinson[2]

[1] Columbus Laboratories, Battelle, Columbus, OH 43201
[2] Office of Toxic Substances, U.S. Environmental Protection Agency, Washington, DC 20460

Compositing selected environmental samples before chemical analysis is a technique used to save analytical costs for estimating population average residue levels of toxic chemicals. In certain situations, compositing can provide greater tissue mass per analytical sample, and thus provide higher levels of analyte for detection of the presence of specified chemicals. A statistical based compositing design is presented for use in a national survey to identify toxic chemicals in human adipose tissue. The sampling design, compositing design, and statistical analysis methods are presented and discussed.

The National Human Adipose Tissue Survey (NHATS) (1) is one of two main operative programs of the National Human Monitoring Program (NHMP). The NHMP is an ongoing chemical monitoring network designed to fulfill the human and environmental monitoring mandates of both the Federal Insecticide, Fungicide, and Rodenticide Act (FIFRA) as amended, and the Toxic Substances Control Act (TSCA).

The general purpose of the National Human Adipose Tissue Survey is the detection and quantification of the prevalences of selected toxic substances in the general U.S. population. The specific objectives of the survey are:
1. To measure average concentrations and prevalences of toxic substances in the adipose tissue of the general U.S. population;
2. To measure time trends of these concentrations;
3. To assess the effects of regulatory actions; and
4. To provide baseline data.

The data needed to meet these objectives are generated on a annual basis by collecting and chemically analyzing adipose tissue specimens for selected toxic substances, mainly organochlorine compounds and polychlorinated biphenyls (PCBs). The 20 compounds that are currently monitored in the study are listed in Table I.

0097–6156/85/0292–0174$06.00/0

Table I. Compounds Monitored in the National Human
Adipose Tissue Survey

p̲,p̲'-DDT	Aldrin
o̲,p̲'-DDT	Dieldrin
p̲,p̲'-DDE	Endrin
o̲,p̲'-DDE	Heptachlor
p̲,p̲'-DDD	Heptachlor epoxide
o̲,p̲'-DDD	PCB
α-BHC	Oxychlordane
β-BHC	Mirex (Dechlorane)
γ-BHC (Lindane)	trans-Nonachlor
δ-BHC	Hexachlorobenzene

The adipose tissue specimens are obtained from a sampling population of surgical patients and autopsied cadavers. A nationwide random sample of selected pathologists and medical examiners collect and send to EPA/OTS adipose tissue specimens obtained on a continuing basis throughout the fiscal year. The pathologists and medical examiners also supply EPA with a limited amount of demographic, occupational, and medical information for each specimen. This information allows reporting of residue levels by subpopulations of interest, namely, sex, race, age, and geographic regions.

EPA/OTS is interested in enhancing its capabilities to provide more meaningful and comprehensive measures of the changes in levels of TSCA chemicals in man and the environment. Part of this effort involves an enhancement to the NHATS network to include additional chemicals beyond those currently monitored. An investigation is being conducted to identify other toxic substances in specific chemical classes that are present in the population at detectable levels.

The investigation involves broad scan chemical analyses (BSA) being performed on the Fiscal Year 1982 collection of adipose tissue specimens. The chemical analyses are being performed on composited samples of specimens to increase the probability of detecting existing toxic compounds, and to minimize analysis costs for estimating average residue levels existing in the population. The specimens are composited according to a statistical design. This permits the probability of detecting existing toxic substances to be controlled, and provides for valid comparisons of residue level distributions among geographical regions and certain demographic categories.

The results of the program will be a redesign of the current monitoring program. The new program will include additional toxic chemicals identified during the broad scan analyses, and will likely include additional collection media beyond adipose tissue. The new media to be added to the program will be those body components which are the most efficient for detecting the additional chemicals. One likely media candidate is blood.

This paper addresses the statistical problems and issues involved in the construction of the compositing design and the approach to the statistical analysis of the chemical analytical results.

Sampling Design

The adipose tissue specimens analyzed in this work were collected under the NHATS program during the 1982 fiscal year. A statistically based sampling design was used to select the specimens so that a representative sample could be obtained to make statistical inferences and estimate sampling errors.

A compositing design was developed to provide for control of the analytical errors associated with the broad scan chemical analyses. The inherent limitations of the analytical detection techniques are minimized by increasing the amount of material (and hence analyte) available for analysis. Together the sampling and compositing designs provide for control of the overall error rates associated with the detection of toxic substances. The sampling design is described in this section and the compositing design is described in the next section.

Survey Design. The target population for the NHATS program is the general U.S. population. However, due to practical reasons it is not possible to directly sample this population. The NHATS survey design is therefore based on a sample population of surgical patients and autopsied cadavers. This limitation requires the assumption that the prevalences and levels of the substances of interest are the same in both populations.

Sample Selection. The survey design involves a multistage process to select a nationwide sample of cooperating pathologists. The 48 conterminous states are stratified into the nine Census Divisions. Within each Census Division, Standard Metropoliton Statistical Areas (SMSAs) are selected with probabilities proportional to their population. The number of SMSAs selected within a Census Division is determined by its relative population with respect to the general U.S. population. Within each SMSA a selected pathologist or medical examiner is asked to supply tissue specimens according to a specified quota based on the age, sex, and race category of the specimen. The categories considered are: age (0-14 years, 15-44 years, 45 years and older), sex (male, female), and race (white, nonwhite). Each quota is based on the age, sex, and race distributions of the associated Census Division. Within each SMSA the specimens are selected in a nonprobabilistic manner based on the judgement of the professionals involved.

An overview of the sampling stages is presented in Figure 1. The geographic stratification ensures a representative sample from all regions of the country, and improves the ability to make regional and national estimates of prevalences and levels of the toxic substances. Quotas specified for each age, sex, and race category ensure the proper representation for these different groups.

Sample Weights. The true probability of selection for specimens collected in the NHATS network cannot be calculated since some stages of the sample selection involve nonprobability sampling. For example, the specimens are selected by the pathologist in a non-statistical manner. In addition, the sample is not self-weighting due to discrepancies between actual samples and design quotas. Therefore, sample weights are assigned to the

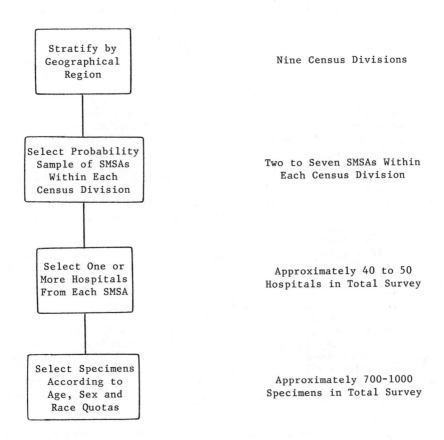

Figure 1.   Overview of the Sample Selection Process

specimens so that bias is minimized for estimates of regional and national averages.

Since the SMSAs selected represent a valid probability sample, the weight assigned to each specimen is based on the selection protocol. Various adjustments are performed on the weights so that the sum of the sample weights for a given Census Division, age, sex, and race category equals the corresponding population count according to the U.S. Census. The weight assigned to an individual specimen therefore reflects the number of individuals in the general population represented by that specimen.

Compositing Design

The objectives for compositing tissue specimens before chemical analysis are:
1. To increase the amount of sample material and analyte in the sample so that the probability of detection is increased; and
2. To obtain a tissue sample more representative of the average toxic residue level existing in the population in order to reduce chemical analysis costs required to achieve a specified precision of an estimate.
The compositing design must address these two objectives.

The number of individual specimens assigned to each composite is determined by:
● The sensitivity of the analytical instrumentation; and
● The magnitude of the population residue levels that are of interest to be detected.
The details concerning how these factors determine the required number (N) of specimens per composite are described in the following sections.

Sensitivity of the Analytical Instrumentation. Prior to the chemical analyses for each compound, a series of calibration tests are run to determine the relationship between instrument response and the true analyte level in a sample. The form of the calibration curve and the instrument response variability about this curve determine the analytical sensitivity. Two quantities are calculated from the calibration data:
● Limit of detection (LOD) - a threshold value unlikely to be exceeded by the instrument response when no analyte is present in the sample.
● Minimum detectable analyte level (MDL) - the minimum analyte level required in a sample to obtain a high probability of detection.
Responses that exceed the LOD are highly indicative of analyte being present in the sample. The presence of analyte in the sample is therefore declared to have been detected when such exceedances occur.

The MDL represents an analyte level for which an exceedance is almost guaranteed to occur. One objective of the compositing design is to ensure that the analyte level per injection of an aliquot of the composited sample into the analytical instrument exceeds the MDL whenever the population of specimen residue levels exceeds stated levels of interest to be detected.

In the Broad Scan Analysis program, evaluation of the sensitivity of the analytical procedures indicated that an analyte level of 10 ng per injection (i.e. MDL = 10 ng) was sufficient for detection of most compounds of interest.

Population Residue Levels to be Detected. The number of individual specimens required per composite was determined by the MDL values for the different compounds addressed by the broad scan analyses. As indicated previously, a universal MDL of 10 ng was considered to be appropriate for the compounds of interest. The number of individual specimens per composite was chosen so that there was a high probability that an injection drawn from the extract of the composited sample contained an analyte level exceeding 10 ng for each compound in which the population residue levels exceed the stated levels that were of interest to be detected. The population residue levels of interest correspond to concentrations expected to be toxic to humans yet are not chosen to be too small to require a very large number of specimens per composite.

Since the population residue levels for each compound actually represent a distribution of values, the stated concentrations that are of interest to be detected correspond to a specification of various parameters of the population distribution. The concentrations are assumed to follow a log-normal distribution. A specific distribution in the log-normal (2) family is determined by two parameters, and thus two characterizations were required to specify the concentrations to be detected by the broad scan analyses. These characterizations involved population average concentrations and standard deviations.

Calculation of N. The composited samples are formed by combining a specified fixed mass of tissue (e.g., 1 gram) from each specimen. The entire composited sample is then extracted down to a prespecified final extract volume. Each analytical determination involves an injection of a small amount of the extract into the analytical instrument.

Let X denote the amount of analyte in a tissue sample of mass w grams taken from a randomly selected specimen in the population. The required number of specimens per composite for a particular compound is the minimum value of N satisfying

$$P(X_1 + \ldots + X_N \geq k \cdot MDL) \geq 0.99 \qquad (1)$$

where k is the ratio of the final extract volume to the volume per injection. For example, if the final extract volume is 50 µL and the volume per injection is 2 µL, then k is 25.

The number of specimens required per composite was the minimum value of N satisfying Equation 1 for all compounds addressed by the broad scan analyses. One constraint on the value of N was that the total mass of the composited sample could not exceed sample preparation and extraction constraints (approximately 30 gram). To actually determine the value of N it was noted that the value of N satisfying Equation 1 depends on the assumed distribution of population residue levels. Various lognormal distributions (i.e. values of the mean and standard deviation) were therefore investigated to determine how N varies with these parameters.

An N value of 20 was ultimately chosen because it satisfied Equation 1 for the type of population distribution residue levels of interest in the broad scan analysis.

Compositing Scheme. The manner in which the specimens are composited is determined by the subpopulations to be compared. Geographical comparisons were the primary interest among the stratification variables, with age, sex, and race being secondary factors of interest. An evaluation of the 1982 specimen set indicated that the best compositing scheme involved compositing specimens within Census Division and age category. Age was chosen as an additional stratification variable since age was considered the most likely other variable to be significant. Further stratification on sex or race was not possible in general, due to the limited number of available specimens. Within a Census Division and age category, composites were therefore formed with varying proportions of males and females, whites and non-whites. This allowed the effects of race and sex to be assessed by comparing composites having different proportions.

Statistical Analysis

One of the objectives of the Broad Scan program was to make comparisons of residue level distributions across geographic regions and, if possible, certain demographic variables. This required the selection of an appropriate statistical model and approach to the analysis. (3)

The Statistical Model. The residue levels of the individual specimens in a particular subpopulation (e.g., a given Census Division and age, sex, race category) are assumed to follow a lognormal distribution. Previous studies on NHATS data have found the lognormal distribution to be appropriate and goodness of fit tests performed on the collected data verified that the assumption is still reasonable. The lognormal model assumes only non-negative values and allows the variances of the different subpopulation distributions to increase with the mean levels. This distribution is commonly used to model pollutant levels in the environment.

The model that describes the residue level of an individual (M gram) specimen sample is given by

$$C_{ijhtk} = M \cdot \mu \cdot D_i \cdot A_j \cdot 10^{\beta_1 RC_h} \cdot 10^{\beta_2 SX_t} \cdot S_{k(ijht)} \qquad (2)$$

where

μ represents the overall population average residue level;

D_i represents the effect (i.e., the deviation from μ) due to the ith Census Division;

A_j represents the effect due to the jth age category;

$RC_h = \begin{cases} 1, \text{ if the specimen is White} \\ -1, \text{ if non-White}; \end{cases}$

$SX_t = \begin{cases} 1, \text{ if the specimen is Male} \\ -1, \text{ if Female}; \end{cases}$

β_1 and β_2 are unknown coefficients to be estimated from the data; and $S_{k(ijht)}$ represents the effect of the kth specimen randomly selected from the given age, sex, race and Census Division category.

The effect due to specimen differences is assumed to be a random variable having a lognormal distribution. The assumption of a lognormal distribution implies that the logarithm (base 10) of the random variable has a normal distribution with some mean μ and variance σ^2 [i.e., $\log X \sim N(\mu, \sigma^2)$]. Here we assume that

$$\log S_{k(ijht)} \sim N(0, \sigma^2)$$

where σ^2 denotes the variation associated with the different specimens within a subpopulation.

Compositing involves the summation (or equivalently, the averaging) of residue levels over a number of different specimens. The statistical model for a composite sample is therefore obtained by averaging, over the different specimens involved, the model given in Equation 2. This composite model, however, is very complicated and cannot be directly analyzed by existing statistical theory. An approximate model was therefore used. The approximate model assumes that the experimental effects are additive on a logarithmic (base 10) scale. For a composite sample consisting of N specimens (each of M grams) the model is given by:

$$
\begin{aligned}
C^*_{ij\ell} &= \log \frac{C_{ij\ell}}{M \cdot N} \approx \log \mu + \log D_i + \log A_j \\
&+ \beta_1 \cdot \left(\frac{(N_{11} + N_{12}) - (N_{21} + N_{22})}{N} \right) \\
&+ \beta_2 \cdot \left(\frac{(N_{11} + N_{21}) - (N_{12} + N_{22})}{N} \right) \\
&+ \log S_{\ell(ij)} + \log E_{\ell(ij)}
\end{aligned}
\tag{3}
$$

where

$C_{ij\ell}$ is the residue level for the ℓth composite from the ith Census Division and jth age group;

$S_{\ell(ij)}$ is the unique effect due to the specific specimens comprising the composite;

$E_{\ell(ij)}$ represents the variation due to measurement error; and

N_{ht} equals the number of specimens in the composite having race h and sex t, with $N = N_{11} + N_{12} + N_{21} + N_{22}$.

The multiplier for the coefficient β_1 in Equation 3 is:

P_1 = (Proportion of Whites in the composite)
   - (Proportion of non-Whites in the composite).

If the composite consists of all white specimens, then the race effect is β_1. If the composite consists of all non-white specimens, then the race effect is $-\beta_1$. Thus, β_1 represents the effect due to race. Similarly, β_2 represents the sex effect.

Note that the model given in Equation 3 describes a composite
sample where each specimen contributes M grams. Due to differences
in the number of available specimens for certain composites, the
value of M was varied across some composites. For those composites
where the available number of specimens, N, was small (e.g. N=5),
M was increased accordingly so that M·N remained approximately
constant. This produced composites having approximately the same
amount of mass and hence the same expected average level of analyte.
However, the estimation of parameters in Equation 3 now required
use of weighted regression since composites having larger N values
have smaller variances and thus deserved more weight in the analysis.
Note that the variable $C_{ij\ell}/M \cdot N$ in Equation 3 represents the amount
of analyte on a "per gram" basis.

Comparison of Subpopulations. The comparison of subpopulations
are done on two bases:
- The number of sample composites in a subpopulation exhibiting
  the presence or absence of each toxic substance; and
- Comparison of various parameters (e.g. mean, standard deviation)
  of the residue level distributions across the subpopulations.
Comparison of the number of composites indicating the presence
of a toxic substance is not done in a formal statistical manner.
Rather, the observed results of the chemical analyses are simply
reported. Summaries are given for the different subpopulations.
The objective of this kind of a comparison is to provide qualitative
information concerning where, if any, high residue levels are
apparent.
      The formal statistical comparison of residue distributions
across the various subpopulations involves estimation of the
parameters in the model given by Equation 3. The model assumes
that each subpopulation distribution is lognormal but possibly
differ in mean residue levels and variances. Significant differences
in Census Divisions correspond to significant differences in the
D_i values. Differences in the A_j's correspond to differences among
the age categories. The coefficients β_1 and β_2 provide information
concerning race and sex differences.
      The statistical model fitting is performed to estimate the
parameters of the model. These estimates then provide information
concerning the influence of each demographic and geographic factor
on concentration level of the composite.
      A second analysis using weights based on population census
figures is also performed so that estimates can be made of the
mean residue levels for the different subpopulations. Each specimen
represents a particular number of individuals in the general
population and these values serve as the sample weights.

Summary

The use of a statistical-based sampling design and compositing
design ensures that the results and inferences made from the broad
scan data are defensible. The sampling design provides for control
of the sampling errors which are due to the fact that the sample
upon which inferences are to be made represents only a subset of
the population of interest. The compositing design provides for
control of the analytical errors so that the precision and

sensitivity of the measurements are sufficient to meet the study's objectives.

There are a number of statistical assumption required in developing sampling and compositing designs. The validity of these assumptions are somewhat subjective. The statistical approach is therefore intended to provide only a guideline and framework for conducting the study. The compositing scheme indicates a "ballpark" number of specimens required per composite to meet the study objectives. The value of N therefore represents a nominal number of specimens per composite to detect the type of residue levels desired.

Acknowledgments

The work presented in this paper was partially funded (Battelle's efforts) by EPA Contract No. 68-01-6721.

Disclaimer

This document has been reviewed and approved for publication by the Office of Toxic Substances, Office of Pesticides & Toxic Substances, U.S. Environmental Protection Agency. The use of trade names or commercial products does not constitute Agency endorsement or recommendation for use.

Literature Cited

1. "The Program Strategy for the National Human Adipose Tissue Survey", U.S. Environmental Protection Agency, draft document, 1984.

2. Johnson, N. L. and Kotz, S. "Continuous Univariate Distributions", John Wiley and Sons: New York, N.Y., 1970, p. 112.

3. Neter, J., Wasserman, W. "Applied Linear Statistical Models"; Richard D. Irwan, Inc.: Homewood, Illinois, 1974, p. 123.

RECEIVED July 17, 1985

The Alpha and Beta of Chemometrics

George T. Flatman and James W. Mullins

Environmental Monitoring Systems Laboratory, U.S. Environmental Protection Agency, Las Vegas, NV 89114

Because of the importance of their decisions and the need for statistical justification of their results, monitoring statisticians and chemometricians are being asked by their customers to use hypothesis testing with its attention to false positives and false negatives. This paper explains the prerequisite assumptions, logic flow, and customary confidence values (alpha, beta) of classical random variable hypothesis testing. An algorithm, equating the expectations of the loss values of a false positive and a false negative, calculates the ratio of alpha to beta given a site specific beta rather than the customary arbitrarily fixed value. Two real-world examples are given to illustrate the extreme variability of estimated beta values. The conclusion states the need for hypothesis testing in monitoring activities and the need for site specific alpha and beta algorithms in hypothesis testing.

Chemometrics and monitoring statistics often are used to make very exacting decisions with potentially costly and contested consequences. Conclusions are presented with statistical textbook vocabulary but not always with statistical reliability. A statistically significant difference may suggest the presence of pollution or suggest only the underestimated variance or skewness of the distribution of the test statistic. In hypothesis testing, this latter case is called a false positive and its probability is called alpha. The power of the test to detect clean as clean is one minus alpha. A sustained null hypothesis may suggest no pollution or suggest only the sample size was too small. In hypothesis testing, this latter case is called a false negative and its probability is called beta. The power of the test to detect polluted as polluted is one minus beta. Chemometrics, like monitoring statistics, needs to use all of hypothesis testing. All include alpha, beta, power to detect clean as clean (1-alpha) and power to detect dirty as dirty (1-beta).

Monitoring statistics starts with a random variable design to find and describe a toxic chemical site by a <u>mean</u> and a <u>variance</u>. If a large and intense plume is found, then geostatistics is used to find the structural pattern of the toxic substance in time and/or space. If a strong correlation structure exists, then monitoring statistics can draw a contour map of the toxic substance plume by means of spatial variable methods such as Kriging. As the environmental scientists calculate means, variances, and contour maps, the risk assessors and health scientists need to know how good are these statistics. They are asking what alpha (probability of calling clean "polluted"), what beta (probability of calling polluted "clean"), and what powers of the test (1-alpha, the probability of calling clean "clean" and 1-beta, the probability of calling polluted "polluted") do the site selection criteria or clean-up criteria have. In response to these questions, monitoring statistics and chemometrics must apply meaningfully the statistical abstractions "alpha," "beta," and "powers of the tests." These are well defined for random variables. This presentation discusses especially the beta-problems plaguing monitoring statistics in random variable hypothesis testing. The U.S. EPA's Environmental Monitoring Systems Laboratory-Las Vegas is extending "alpha," "beta," and "powers" to spatial statistics. The task is complicated by the shifts from single inference to multiple inference and from random variable to spatial variable.

The logic of the hypotheses testing was developed by R. A. Fisher for the needs of the agricultural experiment station. The logic is simple and obvious but should be worked out carefully step by step. In the rush of the workaday world, overworked scientists often fail to think through clearly the hypotheses which they are testing. This can lead to a powerless experiment that proves only that the number of samples taken was too small.

First the hypotheses must be chosen. There are two: (1) the null hypothesis denoted by H sub zero which is assumed true until rejected, and (2) the alternative hypothesis denoted by H sub one or sub A for alternative which is assumed false until the null hypothesis is rejected. The logic of the test requires that the hypotheses be "mutually exclusive" and "jointly exhaustive." "Mutually exclusive" means that one and only one of the hypotheses can be true; "jointly exhaustive" means that one or the other of the hypotheses must be true. Both cannot be false. The null hypothesis is to reflect the status quo, which means that failure to reject it is only continuation of a present loss. For the agricultural station, failure to improve the status quo means that the old brand of seed, pesticide, or fertilizer is used when, in fact, a new and better brand is available. This is a status quo loss of productivity (e.g. 10 percent lower yield), but one which the farmer unknowingly accepts. The loss from a customary alternative hypothesis might be 90 percent of the crop destroyed by disease or insects that the old strain was immune to, or a new fertilizer or pesticide that is found to leave a carcinogenic residue exceeding an action level in 90 percent of the crop. Obviously, the status quo loss is smaller in the extreme case than the potential alternative loss. Management science statistics

often uses worst case expected losses in evaluating alternatives.
If the decision maker can tolerate the worst case loss, then he can
use that alternative. The expected value of the losses will be very
important in the discussion of beta. Will the loss from calling a
polluted area clean be minimal among the losses associated with the
tests of pollution hypotheses?

For monitoring toxic substances, such as dioxin cleanup, assume
we have calculated an $\bar{x}$ and s for each unit area or rectangular panel
potentially needing cleanup and have been given an action level of 1
ppb. The action level is a constant and has no variance. The $\bar{x}$ and
s are computed from a field triplicate of a composite of subsamples
equally spaced from a uniform grid covering the panel. The null
hypothesis says "no difference," and represents the status quo.
Hopefully, nonpolluted or less than 1 ppb is the status quo, and polluted or equal to or larger than 1 ppb is the exception.

Let $\bar{x}_i$ be the mean in ppb from panel i

s_i be the standard deviation in ppb from panel i

Null Hypothesis: This panel is clean

Ho: $\bar{x}_i < 1$ ppb

Alternative Hypothesis: This panel is polluted

Ha: $\bar{x}_i \geq 1$ ppb

Let: TS be a test statistic which approximates a Student's
t-distribution

$_{df}t_\alpha$ be table value of t-distribution for appropriate
degrees of freedom (df), alpha (α), confidence level
for a one tail test

CV be the critical value of $_{df}t_\alpha$ from the t-table.

df = 3 - 1 or 2

α = .05

$$TS = \frac{\bar{x}_i - 1}{\dfrac{s_i}{\sqrt{3}}}$$

CV = $_{df}t_\alpha$

If (TS < CV), there is no reason to reject the null hypothesis
and if (TS $\geq$ CV), the null hypothesis is rejected, implying the
alternative hypothesis is true.

In Figure 1 the decision space is represented by the bottom
horizontal line which is divided by the vertical line representing
the critical value (CV). The segment of the line less than CV represents the part of the decision space sustaining the null hypothesis.
The segment of the line equal to or greater than CV represents the
part of the decision space rejecting the null hypothesis and accept-
ing the alternative hypothesis. The upper horizontal line represents
the real line and the value of the test statistic (TS). Again the
line is divided by the value of CV. The height or ordinate of the
curve represents the probability that the test statistic (TS) takes
on the value of abscissa. The bell shape of the curve shows that
TS has a high probability of taking the abscissa values near the
center of each distribution (Ho or Ha) and a low probability of
taking the values in the tails. The dashed shaded area represents
the distribution of the TS under the null hypothesis, and the dotted
shaded area represents the distribution of the TS under the alternative hypothesis. Classical statistics assumes identically distributed and equal variances; therefore, the shaded areas are the
same shape with equal spreads but different locations (different
means). Note the decision space (bottom line) is discrete but the
"real world" data of the real line and shaded distributions overlap.
This overlap gives rise to the possibility of error labeled in Figure
1 as alpha, a dashed area right of CV, and beta, a dotted area left
of CV. Alpha and beta appear equal in Figure 1. Their relative
size is the concern of this paper. Since the area (cumulative
probability) of a probability distribution must add to one, the area
of no error (correct decision) is represented by the dashed area
below CV one-minus-alpha and the dotted area above CV one-minus-beta.

Then:

Alpha (α) is the probability of calling a clean panel
polluted or the type I error and shown as dashed
area to the right of CV in Figure 1.

Beta (β) is the probability of calling a polluted panel
clean or the type II error and shown as dotted
area to the left of CV in Figure 1.

One-minus-alpha ($1 - \alpha$) is the probability of calling a
clean panel clean and shown as dashed area to
the left of CV in Figure 1.

One minus beta ($1 - \beta$) is the probability of calling a
polluted panel polluted, is called
the power of the test, and is shown
as dotted area to the right of CV in
Figure 1.

Now the conventional value for alpha is .05, giving .95 probability of calling a clean panel clean. The probability of beta (β)
depends on the true value of the mean in the alternative distribution
and on the testing assumptions of: (1) equal standard deviations and
(2) identical frequency distributions. Since the difference between

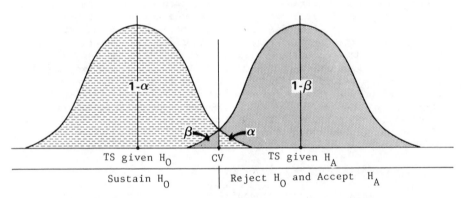

Figure 1. The lower line represents the discrete decision space, the upper line represents the real values that the test statistic (TS) may take, and the overlapping shaded areas represent the probability that the test statistic takes these real values under each hypothesis.

the null and alternative hypotheses is the difference between the
dispersion of no pollutant and the dispersion of a pollutant, it
seems reasonable that there would be a different standard deviation
and frequency distribution, thus contradicting the assumptions of
hypothesis testing; however, answering this problem is beyond the
scope of this paper. Assuming equal alternative standard deviation
and distribution, an acceptable beta (β) has been classically set in
USDA's USFS Experiment Station work at .20 or less. However, this
beta, four-times-larger than alpha, is based on the assumption that
type II error has lower loss value. What are the loss values of:
(1) cleaning a panel that is already clean (type I) and (2) leaving
dirty a panel that is in fact polluted (type II)? Management science
statistics uses expected loss to make probabilistic losses comparable.

$$E(LOSS) \quad = \quad \text{probability of loss x value of loss}$$

For type I: $E(LOSS)$ = α x value of loss from committing a type
I error

For type II: $E(LOSS)$ = β x value of loss from committing a type
II error.

With a fixed sample size, the magnitudes of α and β are inversely
related; that is, if alpha decreases by moving the CV to the left
then beta increases, and if alpha increases, then beta decreases.

Increasing the sample size would reduce both alpha and beta, but
samples and especially their analyses cost money. Intuitively the
minimal actual loss should occur when the expected losses are equal.
So the relative alpha and beta should be found from equating expected
loss from type I error with the expected loss from type II error.

$$E \text{ (type I loss)} \quad = \quad E \text{ (type II loss)}$$

$$\alpha \text{ x (loss from type I error)} \quad = \quad \beta \text{ x (loss from type II error)}$$

$$\beta : \alpha :: \text{ (loss from type I error):(loss from type II error)}$$

For example, in a soil cleanup, the loss from type I error or cleaning a clean panel might be the cost of scraping up six inches of soil
within the panel, trucking the soil away, and disposing of the soil;
probably a cost measured in hundreds to thousands of dollars. The
loss from a type II error or leaving a polluted panel would have a
wide range of potential costs from nothing to the adverse human
health effects. I suggest that realistically the health effects'
cost is at least as high as the cost of the unneeded cleanup, or in
the magnitude of hundreds to thousands of dollars. Mathematically
this means:

$$\beta : \alpha :: \text{ (loss from type I error):(loss from type II error)}$$

$$\beta : \alpha :: \text{ (cost of cleaning a panel):(cost of human's health)}$$

$$\beta : \alpha :: \text{(hundreds of dollars)}:\text{(hundreds of dollars)}$$

$$\Longrightarrow \beta = \alpha$$

Note that if beta equals alpha, beta is one fourth of the tradition-
ally allowed type II error (i.e., .05 instead of .20). This shows
that the unthinking use of textbook examples or traditional confidence levels can be dangerous to the environment and public health.
Pollution monitoring statistics must have its own beta calculations.

Next apply this analysis of the hypotheses testing logic to the
proposed monitoring of a RCRA dump site using the Fisher-Behrens Test
(another Student t-distribution). If a clean ground-water sample is
diagnosed as polluted (type I error), the corrective action is resam-
pling and reanalysis which would cost a few hundred dollars, but
diagnosing a polluted ground-water sample as clean (type II error)
may allow a ground-water pollution plume to grow to a size that will
require a cleanup of thousands or tens of thousands of dollars.
Mathematically this means:

$$\beta : \alpha :: \text{(loss from type I error)}:\text{(loss from type II error)}$$

$$\beta : \alpha :: \text{(cost of resampling and analysis)}:\text{(cost of}$$
$$\text{ground-water cleanup)}$$

$$\beta : \alpha :: \text{(hundreds)}:\text{(tens of thousands)}$$

$$\beta : \alpha :: 1 : 100$$

Note that in this case, beta should be one one-hundredth of alpha.
Again the unthinking use of textbook examples or traditional confidence levels can be dangerous to the environment and public health.
Even the previously calculated beta for the soil cleanup example is
two orders of magnitude too large.

In conclusion, chemometrics, like monitoring statistics, requires an alpha and beta which differ from classical values. Especially beta must be calculated by statistical expectations for each
application. Conventional values of beta or values of a previous
pollution site may be incorrect by orders of magnitude for the current site. Statistics is not a tool that can be used by rote; thorough understanding and site-specific thought is essential. The alpha
and beta of monitoring statistics is site-specific. If alpha is an
acceptable type I error for the test, then one minus alpha is an
acceptable power for calling clean "clean," and if beta is an acceptable type II error for the test, then one minus beta is an acceptable
power for calling polluted "polluted." All four values must be
thought out.

RECEIVED July 17, 1985

Statistical Issues in Measuring Airborne Asbestos Levels Following an Abatement Program

Jean Chesson[1], Bertram P. Price[2], Cindy R. Stroup[3], and Joseph J. Breen[3]

[1] Columbus Laboratories, Battelle, Columbus, OH 43201
[2] National Economic Research Associates, Inc., Washington, DC 20036
[3] Office of Toxic Substances, U.S. Environmental Protection Agency, Washington, DC 20460

Asbestos abatement activity, especially removal projects in schools and other public buildings, is ex-
panding rapidly in response to EPA's recent "Asbestos
in Schools" rule. A focal point for the many
technical and scientific problems in EPA's asbestos
program is the question of how to determine that a
contractor has successfully completed an abatement
project. The question engulfs the broad debate on
biologically effective fibers, sampling strategies and
analytical alternatives. For any post-abatement
evaluation protocol that is proposed, there are
important statistical design and analysis issues that
must be addressed. The statistical objectives are to
obtain precise and accurate estimates of airborne
asbestos levels and to quantify and control the
likelihood of a false positive or false negative test
result. In this paper the statistical design issues
are identified and the relationship between sample
size, error rates and cost is analyzed.

EPA has not set an exposure standard for asbestos. However, the
"Friable Asbestos Containing Materials in Schools, Identification
and Notification Rule" was promulgated to address the problem of
exposure for young people whose risk is increased if only because of
their longer expected life span relative to the latency period associated with asbestos-related disease. The rule requires inspection
and testing for asbestos, and notification of maintenance workers,
teachers and parent groups if asbestos is found. The rule does not
require abatement. If asbestos is present, the choice of an
approach and method, if any, for reducing exposure is left to the
local decision-making body.

A building owner who decides on abatement is confronted with a
variety of technical and scientific issues directly affecting the
quality of the job to be undertaken. EPA guidance identifies these

issues and recommends that a program manager be designated to coordinate the activities of contractors and consultants. One of the
consultant's most important functions is to ascertain that the project has been successfully completed and that the contractor can be
released.

EPA guidance on this question--often referred to as the "how
clean is clean" issue--has been cautious. The Agency has been careful not to recommend post-abatement evaluation techniques that have
not been thoroughly validated. In Guidance for Controlling Asbestos
Containing Materials in Buildings (1), EPA offers a recommendation
to assess whether a contractor has effectively reduced the elevated
airborne concentration levels generated while the work was in
progress. A visual inspection is to be conducted to ascertain that
all asbestos-containing materials have been removed and no debris or
dust remains. Air sampling is also recommended. A slow, long
sample (2 liters per minute for eight hours) should be taken within
48 hours after the work has been completed. It is recommended that
the samples be analyzed by Phase Contrast Microscopy (PCM).
Although PCM is only sensitive to large fibers (may miss long thin
fibers) and does not distinguish asbestos from other types of
fibers, other analytical methods based on electron microscopy that
may be definitive had not been sufficiently validated to receive a
recommendation at the time the guidance document was prepared.
Furthermore, PCM is satisfactory for its intended purpose, namely,
to determine if an asbestos worksite, the abatement area, has been
restored to its normal condition.

As a result of a recent EPA/NBS workshop on "Monitoring and
Evaluation of Airborne Asbestos Levels Following an Abatement
Program," many of the apparent conflicts among sampling and
analytical methods were tempered. A variety of new proposals involving both optical microscopy and electron microscopy appeared to
be acceptable for post-abatement monitoring. The workshop did not
explicitly address questions of uncertainty in the measurement process, the likelihood of reaching a correct conclusion or implementation cost. We address these questions using TEM as an example.
With slight modification the approach and solutions obtained are
applicable to other sampling and analysis methods.

Statistical Issues

The objective in measuring post-abatement airborne asbestos concentrations is to confirm that the levels in the area of interest do
not constitute a public health hazard. To date EPA has not promulgated a numerical exposure standard for airborne asbestos. In
general, the scientific information regarding health effects related
to exposure to asbestos at environmental levels is currently insufficient to serve as a basis for standard setting. However, asbestos
abatement projects are ongoing and the parties engaged in this work
need criteria to determine when a contractor has fulfilled the
requirements. In the absence of an absolute standard, one feasible
approach to releasing a contractor is to base the decision on the
comparison of two requirements.

We propose a criterion based on a comparison between the post-
abatement indoor airborne asbestos level and the ambient (outdoor)
level. An area is "clean" when there is no statistically

significant difference between the indoor and ambient levels. This
approach is not currently under formal consideration for rulemaking.
   Once a criterion is accepted it is possible to determine an
optimal scheme for collecting the necessary data. There are two
broad issues that must be addressed--air sampling protocols including quality assurance, and statistical sampling designs. Determining the best air sampling protocol involves decisions about the type
of sampling to be used (e.g., whether or not to employ aggressive
sampling which involves stirring up the air), flow rates, duration,
frequency and type of filter. Some of these choices are tied to the
type of analysis--PCM, SEM, or TEM--that is to be done. Other
choices, for example, rate, duration, and frequency are sample size
issues which depend on statistical requirements such as false
positive rate, false negative rate and sensitivity. Sensitivity in
this context means the magnitude of difference between indoor and
outdoor or pre and post levels that must be detected with a high
probability. A sampling and analysis program should have sufficient
sensitivity to detect the smallest difference between indoor and
ambient levels that is considered to be unacceptable.

Statistical Model. In designing an airborne asbestos sampling
analysis procedure for determining if an area is clean, the physical
sampling parameters and the statistical design parameters are inter-
related. For example, static area sampling and aggressive local
sampling may dictate different statistical designs. The volume of
air sampled--duration multiplied by sampling rate--affects detection
limits and variability. Characteristics of the area, its size,
whether it is one contiguous space or partitioned into unique sections (rooms), and other distinguishing factors are important. However, for our current purpose, we have simplified the discussion.
We consider an area the size of a typical room. We can then focus
on the generic sampling problem and isolate the relevant statistical
issues.
   A typical measurement may be described as

$$Y_{ijk} = lnZ_{ijk} = \mu + T_i + L_{j(i)} + \varepsilon_{k(ij)} \qquad (1)$$

where

Z_{ijk} is the kth laboratory analysis taken at the jth location in the
i^{th} time period. (Typically Z_{ijk} will have units of ng/m^3 or
fibers/m^3. The discussion that follows does not depend on the
choice of units.)
   T_i, $L_{j(i)}$ and $\varepsilon_{k(ij)}$ are random variables representing the
contribution of time period, location and analytical error,
respectively, with zero means and variances σ_T^2, σ_L^2 and σ_E^2. The
model represents a completely nested design. We have implicitly
assumed that the measurements have a lognormal distribution. Under
this assumption the natural logarithm of the analytical measurement
has a variance that does not vary with its mean and standard
statistical tests can be applied.
   The variance of an individual measurement is

$$Var(Y_{ijk}) = \sigma_T^2 + \sigma_L^2 + \sigma_E^2 \qquad (2)$$

The variance of the average taken over laboratory analyses, locations, and time is

$$\text{Var } (\bar{Y} \ldots) = \sigma_T^2/I + \sigma_L^2/IJ + \sigma_E^2/IJK \tag{3}$$

where I, IJ, and IJK represent the total number of time periods, locations and laboratory analyses, respectively. Since the purpose of sampling is to determine, as quickly as possible, if the airborne asbestos levels are low enough to release the contractor, we restrict the discussion to values of I = 1, that is, a single time interval.

Sample Allocation for Estimating Concentration Levels. The variable cost of a sampling program that produces an estimate with variance given in equation (3) is

$$C = C_1*J + C_2*JK \tag{4}$$

C_1 is the unit cost of collecting a sample and C_2 is the analysis cost.

The sample size and sample allocation scheme is obtained in one of two ways. Either the cost is fixed and the variance of the mean is minimized or the variance is fixed and the cost is minimized. The first approach is used when the budget for sampling and analysis is determined in advance. The objective in this case is to use that budget to obtain an estimate with maximum precision (equivalently minimum variance). The second approach is used when the required precision of the estimator is specified in advance. Then the objective is to derive an estimator with the desired level of precision at the lowest possible cost.

Minimum Variance, Fixed Cost. The mathematical problem is

$$\min_{J,K} \quad \sigma_L^2/J + \sigma_E^2/JK \tag{5}$$

subject to

$$C_1 J + C_2 JK = C \tag{6}$$

where C is the fixed dollar total available for the sampling and laboratory analysis portion of the project. The solution is

$$J = C/[(C_1C_2)^{\frac{1}{2}} (\sigma_E/\sigma_L) + C_1]$$

$$K = (C_1/C_2)^{\frac{1}{2}} (\sigma_E/\sigma_L) \tag{7}$$

To develop a feeling for the sensitivity of the allocation scheme to values of σ_L and σ_E, we have prepared examples that reflect passive sampling followed by TEM analysis. We use values of C_1 = \$20 and C_2 = \$500 and evaluate the solution for total budgets of \$5,000 and \$15,000. These figures are based on typical costs for TEM analysis. Costs will vary among laboratories and depend on the analytical method (TEM, SEM, or PCM) chosen. Note that σ_L and σ_E

are needed to compute the level of precision achieved, namely, (Var
Y ...) Table I   shows results for σ_L = .25, .5, 1.0, 1.5, and 2.0
and for σ_E = 1.0 and 1.5. Results calculated from several studies
suggest that total variation (σ_L^2 + σ_E^2) ranges between 1.0 and 2.0.
See (2-4).

Table I.   Solution to the Sample Allocation
Problem: Minimum Variance, Fixed Cost

Values of σ_L	Nominal Cost	Values of σ_E								
		1.0					1.5			
		σ_E/σ_L	J	K	Error Bound %	Actual Cost	σ_E/σ_L	J	K	Error Bound %	Actual Cost
.25	5,000	4.00	10	1	89.4	4,950	6.00	5	2	159.9	4,850
	15,000	4.00	29	1	45.5	14,830	6.00	15	2	73.6	15,050
.50	5,000	2.00	10	1	100.0	4,950	3.00	10	1	166.4	4,950
	15,000	2.00	29	1	50.2	14,830	3.00	29	1	77.8	14,830
1.00	5,000	1.00	10	1	140.3	4,950	1.50	10	1	205.7	4,950
	15,000	1.00	29	1	67.3	14,830	1.50	29	1	92.7	14,830
1.50	5,000	0.67	10	1	205.7	4,950	1.00	10	1	272.4	4,950
	15,000	0.67	29	1	92.7	14,830	1.00	29	1	116.4	14,830
2.0	5,000	0.50	10	1	299.8	4,950	0.75	10	1	370.9	4,950
	15,000	0.50	29	1	125.6	14,830	0.75	29	1	148.4	14,830

Solutions to Equations 7 produce solutions for K and J that are
not whole numbers. To resolve this problem, we increase K to the
next integer value and choose J so the cost constraint is exceeded
by the smallest amount possible.
     For the majority of cases in the table, only one laboratory
analysis of the filter is required, K = 1. This happens because the
cost of analysis is large relative to the cost of collecting an additional sample. The value of K will be larger than 1 if the
analytical variation, σ_E, is larger than the spatial variation, σ_L,
by enough to overcome cost disparity. For our example, the ratio of
σ_E to σ_L must be greater than 5 to cause K to be greater than 1.
     Although the values obtained for J and K minimize the variance,
we gain more insight into the meaning of the numbers in Table I by
describing them in terms of an error bound for estimating asbestos
level.   A 95% confidence interval for the mean of the log-
transformed data is $\bar{Y} \pm 1.96$ SD($\bar{Y}$). In terms of untransformed data
the confidence bounds are exp($\bar{Y}$ - 1.96 SD($\bar{Y}$)), exp($\bar{Y}$ + 1.96 SD($\bar{Y}$)).
These limits determine a confidence interval for the median of the
untransformed data. The error bound is calculated as

$$e^{1.96SD(\overline{Y})}{-1}$$ (8)

For the case $\sigma_E = 1.0$, $\sigma_L = .25$ and $C = \$5,000$, the upper error bound is 89.4 percent.

This result means we have confidence level of .95 that the actual asbestos level will be no greater than 89.4 percent higher than the estimated level. For the same values of σ_E and σ_L, increasing the budget to $\$15,250$ yields an upper bound of 45.5 percent. That is, an increase in the budget by a factor of 3 can achieve a reduction in estimation error by approximately a factor of 2. A fairly real-istic but conservative case may be $\sigma_E = \sigma_L = 1.0$. In this case, for the $\$5,000$ option, the upper bound is 140 percent above the estimated value. For the $\$15,000$ option, the upper bound is 67 per-cent above the estimated value. As in the previous case, increasing the testing budget by a factor of 3 cuts the estimation error approximately in half.

Minimum Cost, Fixed Variance. Since our interest was focused on estimation error, we may choose to set the error in advance and allocate sampling and analysis resources to minimize cost. The mathematical problem to be solved is

$$\min_{J,K} \quad C_1 J + C_2 JK$$ (9)

subject to

$$\sigma_L^2/J + \sigma_E^2/JK = V$$ (10)

The solution is

$$J = \frac{\sigma_L \sigma_E}{V} \left[\frac{\sigma_L}{\sigma_E} + (C_2/C_1) \right]^{\frac{1}{2}}$$

$$K = (C_1/C_2)^{\frac{1}{2}} \frac{\sigma_E}{\sigma_L}$$ (11)

Table II displays numerical results for values of $\sigma_E = 1.0$ and 1.5 and $\sigma_L = .25$, .50, 1.00, 1.50 and 2.00. Values of J and K are determined for confidence intervals with upper error bound set at 50 percent and 100 percent and confidence coefficients at 95 percent and 99 percent. In general, the table shows that it is very expensive to go from 95 percent to 99 percent confidence. It is also significantly more expensive to obtain an estimate with a 50 percent error than a 100 percent error. For example, if $\sigma_E = \sigma_L = 1.00$, to be 95 percent certain that the error in the estimated level of airborne asbestos is less than 50 percent requires J = 47 and K = 1 at a cost of $\$24,440$. If a 100 percent error in the estimate is acceptable, J is reduced to 16, K remains at 1 and the cost is $\$8,320$.

Table II. Solution to the Sample Allocation Problem:
Fixed Variance, Minimum Cost

σ_L	Nominal Error %	Confidence %	Value of σ_E					
			1.0			1.5		
			J	K	Cost	J	K	Cost
					$			$
.25	50	95	25	1	13,000	28	2	28,560
		99	43	1	22,360	48	2	48,960
	100	95	9	1	4,680	10	2	10,200
		99	15	1	7,800	16	2	16,320
.50	50	95	30	1	15,600	59	1	30,680
		99	51	1	26,520	102	1	53,040
	100	95	10	1	5,200	20	1	10,400
		99	18	1	9,360	35	1	18,200
1.00	50	95	47	1	24,440	76	1	39,520
		99	81	1	41,120	132	1	68,640
	100	95	16	1	8,320	26	1	13,520
		99	28	1	14,560	46	1	23,920
1.50	50	95	76	1	39,520	106	1	55,120
		99	132	1	68,640	183	1	95,160
	100	95	26	1	13,520	36	1	18,720
		99	46	1	23,920	63	1	32,760
2.00	50	95	117	1	60,840	147	1	76,440
		99	203	1	105,560	254	1	132,080
	100	95	40	1	20,800	50	1	26,000
		99	70	1	36,400	87	1	45,240

It should be noted that estimation errors of 50 percent or 100 percent are not unduly large for airborne asbestos measurements. For example, the estimate of a typical ambient concentration level may be 2 nanograms per cubic meter. The bounds for a 100 percent error would be 1 and 4. Although there is no absolute health standard, it is doubtful if a change from 2 to 4 would affect a decision on whether to release a contractor. That variation would be well within the range of values found in the ambient distribution.

Comparing Two Measurements. The previous discussion introduced the concepts of sample size and sample allocation for estimating airborne concentration levels. We want to apply those concepts to the comparison of two measurements, inside a building versus ambient levels.

Let X and Y denote the two measurements. The models are:

$$X_{jk} = \mu^x + L_j^x + \varepsilon_{k(j)}^x \tag{12}$$

$$Y_{jk} = \mu^y + L_j^y + \varepsilon_{k(j)}^y \tag{13}$$

where X_{jk} and Y_{jk} are the natural logarithms of the kth laboratory analysis taken at the jth location. Let $\bar{X}$ and $\bar{Y}$ represent the average taken over J locations and K laboratory analyses. The difference in means, denoted by $\bar{W} = \bar{Y} - \bar{X}$, is represented by

$$\bar{W} = \mu^y - \mu^x + L_j^y - L_j^x + \varepsilon_{k(j)}^y - \varepsilon_{k(j)}^x \tag{14}$$

with expected value

$$E(\bar{W}) = \mu^y - \mu^x = \delta \tag{15}$$

and variance

$$Var \ (\bar{W}) = 2 \ \frac{\sigma_L^2}{J} + \frac{\sigma_E^2}{JK} \tag{16}$$

For illustrative purposes let K = 1. Then we must determine the sample size, J, that is required to ensure a small probability of making an incorrect decision. Two types of error can occur. The two concentrations may be different (i.e., $\delta \neq 0$), yet we conclude that they are the same (false negative), or the two concentration may be the same (i.e., $\delta = 0$) yet we conclude that they are different (false positive) (see Table III). In the first case the contractor is released even though the work is unsatisfactory. In the second case the area has to be cleaned again even though it was already clean enough. Having specified acceptable rates of false negatives and false positives we can calculate the required values of J and K, given σ_L and σ_E.

Table III. The Types of Error that Can Occur and Their
Probability of Occurrence

| | Actual Situation | |
Conclusion	Indoor Level Less Than or Equal to Ambient Level	Indoor Level Greater Than Ambient Level
Indoor Level Less Than or Equal to Ambient Level	Correct Decision	False Negative (Conclude 'Clean' When It Is Not)
	Probability = 1-α	Probability = β
Indoor Level Greater Than Ambient Level	False Positive (Conclude 'Not Clean' When It Is)	Correct Decision
	Probability = α	Probability = 1-β (Power)
	Significance Level	

Table IV gives values of J when the false positive rate is 5 percent and the false negative rate is either 5 percent or 1 percent. J depends on V and, since K = 1, on the sum $\sigma_L^2 + \sigma_E^2$. A convenient empirical measure of precision in the original units is the coefficient of variation (cv) which is the standard deviation divided by the mean. Note that $\sigma_L^2 + \sigma_E^2 = \ln (1 + cv^2)$. An algebraic difference, δ, on the logarithmic scale translates into a multiplicative difference on the original scale. Table V has been prepared using the multiplicative factors.

For example, if the coefficient of variation is 2 and the level is twice the other then seventy-three samples are required to achieve a false negative rate of 5 percent. To achieve a false negative rate of 1 percent, one hundred and twelve samples would be required. As the acceptable difference between levels increase, the required number of sample required decreases. When one level is one hundred times the other and the coefficient of variation is 1, only three samples are required to achieve a false negative rate of 1 percent.

So far we have assumed that both air levels would have to be determined and therefore that two sets of J samples would have to be collected. In some cases one level may already be available from other records and can be used as a standard of comparison. Table VI shows for this special case how the probability of a false negative depends on the number of samples collected. For small differences between the measured level and the standard a small sample size has an unacceptable high false negative rate. For example, if the mean is five times the standard level and the coefficient of variation is

Table IV. Sample Size (Value of J Given K = 1) Required
to Ensure a False Positive Rate of 5 Percent
and a False Negative Rate of Either 5 Percent
or 1 Percent When Comparing Two Means

Difference Between Means	Coefficient of Variation					
	.75	1	1.5	2	2.5	3
(A) Probability of False Negative = 5 Percent						
2x	23	35	61	72	108	108
5x	5	7	12	16	19	23
10x	4	5	6	8	10	11
100x	2	2	3	3	4	4
(B) Probability of False Negative = 1 Percent						
2x	33	50	89	112	112	112
5x	7	10	17	23	27	33
10x	5	6	8	11	14	15
100x	2	3	4	4	5	5

2 then a sample size of 4 has a false negative rate of 62 percent.
In other words, the contractor will be released in 62 percent of the
cases in which the actual level is five times the standard level.

Summary and Conclusions

Providing a generally acceptable approach to determining "how clean
is clean" for asbestos abatement projects is complicated by many
factors.   First,   there   is   no   absolute   standard specifying   an
acceptable cutoff point for exposure to airborne asbestos.   Second,
there is a number of competing sampling and analysis protocols that
have been proposed.   None have been fully validated.   Finally, data
from completed studies show that statistical sampling and analytic
variability may each be as large as 100 percent (relative to the
estimated concentration level).

However, even against this background of uncertainty and apparent imprecision, in some cases it is possible to measure airborne
asbestos   with   acceptable   precision   through   replication   for   a
reasonable price.   Since there is no exposure standard, "clean" must
be defined by comparing indoor and outdoor levels.   A statistical
comparison of indoor versus outdoor measurements that is significantly different from zero indicates that the indoor space is not
clean.   Tests may be designed that either compare the average of

Table V. Probability of Detecting a Difference Between
a Single Mean and a Standard Level, Given
Sample Size J

Sample Size	Actual Mean Relative to Standard	Coefficient of Variation					
		.75	1	1.5	2	2.5	3
J=2	2x	.19	.15	.12	.11	.10	.10
	5x	.41	.33	.26	.22	.20	.19
	10x	.55	.46	.36	.31	.28	.26
	100x	.87	.78	.65	.58	.53	.50
J=3	2x	.34	.26	.19	.16	.15	.14
	5x	.83	.69	.52	.43	.38	.35
	10x	.97	.90	.75	.65	.58	.53
	100x	1.00	1.00	.99	.98	.96	.93
J=4	2x	.48	.36	.26	.22	.19	.18
	5x	.97	.89	.73	.62	.55	.50
	10x	1.00	.99	.93	.86	.79	.74
	100x	1.00	1.00	1.00	1.00	1.00	1.00
J=5	2x	.61	.46	.32	.27	.23	.21
	5x	1.00	.97	.86	.75	.68	.62
	10x	1.00	1.00	.98	.95	.91	.87
	100x	1.00	1.00	1.00	1.00	1.00	1.00
J=10	2x	.92	.78	.59	.48	.42	.38
	5x	1.00	1.00	1.00	.98	.95	.93
	10x	1.00	1.00	1.00	1.00	1.00	1.00
	100x	1.00	1.00	1.00	1.00	1.00	1.00

a set of indoor measurements with the average of a set of outdoor
measurements or compare the average indoor measurements with a known
value of the outdoor concentration developed from historical data.

For the latter case, comparing the average of a set of indoor
measurements to a known outdoor level, the sample size may be as
small as five. From Table V, we see that 5 measurements are sufficient to detect a ten fold difference between an indoor average and
a known outdoor level with a .91 probability when the coefficient of
variation is 2.5. (Note that if σ_L and σ_E are both equal to 1, the
coefficient of variation is approximately 2.5.)

Larger sample sizes detect smaller differences. For example,
if J = 10, a 5 fold difference could be detected with a .95 probability; a 10 fold difference would almost certainly be detected.
(Refer to Table V).

Differences that are multiples of 5 or 10 appear to be large.
However, in the testing problem being considered they may be acceptable. Ambient outdoor levels of asbestos tend to be low, in the 0-5
ug/m^3 range. A five-fold or even ten-fold increase produces a level
that may still be considered to be within the acceptable range of
the typical outdoor level. Therefore, it may be necessary only to
be able to identify large differences, larger than 5 or 10 times the
outdoor level.

If the outdoor level must be estimated also (i.e., historically determined levels are thought to be inadequate), then a larger number of samples are required. For a coefficient of variation equal to 2.5 and a probability of correctly identifying a 10 fold difference of .95, 10 samples are required indoors and 10 are required outdoors. (Refer to Table IV). To detect a 5 fold increase, a total of 38 samples are required. When both indoor and outdoor levels must be determined, the sampling experiment can be extremely expensive (recall that TEM analysis is approximately $500 per sample).

Reluctance to bear the high cost provides motivation to seek a less expensive solution. Three obvious directions are suggested. First, refine both the sampling and analytical protocol to reduce variation. Method studies in progress should be pursued rigorously to achieve the desired objective. Second, alter the analytical protocol to reduce cost. Third, utilize a sampling plan that is midway between treating the outdoor level as known and treating it as unknown. The use of a small number of outdoor samples, possibly only one, combined with historical data may be sufficient to confirm the outdoor level.

Results on the refinement of protocols for sampling analysis are forthcoming in the report on the EPA/NBS workshop on monitoring. Additional work on statistical design is also in progress to find practical ways to reduce required sample size which in turn leads to reductions in cost and more efficiency in dealing with contractor release after abatement.

Disclaimer

This document has been reviewed and approved for publication by the Office of Toxic Substances, Office of Pesticides & Toxic Substances, U.S. Environmental Protection Agency. The use of trade names or commercial products does not constitute Agency endorsement or recommendation for use.

Literature Cited

1. "Guidance for Controlling Friable Asbestos-Containing Materials in Buildings," U.S. Environmental Protection Agency, 1983, EPA 560/5-83-002.
2. "Airborne Asbestos Levels in Schools," U.S. Environmental Protection Agency, 1983, EPA 560/5-83-003.
3. "Measurement of Asbestos Air Pollution Inside Buildings Sprayed with Asbestos", U.S. Environmental Protection Agency, 1983, EPA 560/13-80-026.
4. Chesson, J.; Margeson, D. P.; Ogden, J.; Reichenbach, N. G.; Bauer, K.; Constant, P. C.; Bergman, F. J.; Rose, D. P.; Atkinson, G. R.; "Evaluation of Asbestos Abatement Techniques, Phase 1: Removal;" Final Report to U.S. Environmental Protection Agency, Contract Nos. 68-01-6721 and 68-02-3938.

RECEIVED July 17, 1985

Estimation of Spatial Patterns and Inventories of Environmental Contaminants Using Kriging

Jeanne C. Simpson

Pacific Northwest Laboratory, Richland, WA 99352

Kriging is a relatively new statistical approach to
spatial estimation. The kriging estimator is a
weighted average which is the "best linear unbiased
estimator." The derivation of the kriging weights
takes into account the proximity of the observations
to the point (or area) of interest, the "structure" of
the observations (the relationship of the squared
differences between pairs of observations and the
intervening distance between them) and any systematic
trend (or drift) in the observations. Additionally,
kriging provides a variance estimate that can be used
to construct a confidence interval for the kriging
estimate. This paper will discuss the assumptions made
in kriging and the derivation of the kriging estimator
and variance. The application of kriging is
demonstrated with lead measurements in soil cores from
two sites near lead smelters and a third site in a
control area.

Kriging (geostatistics) is a relatively new statistical approach to
spatial estimation. Much of the early theoretical work was done by
G. Matheron at the Paris School of Mines in the 1960s. The
development of geostatistics was motivated by D. G. Krige and his
efforts to evaluate the ore in South African gold mines. To this day
most of the research into geostatistical methods is still aimed at
ore and oil reserve estimation. However, in recent years its use has
spread to many other disciplines including sea-floor mapping (1),
hydrologic parameter estimation (2), ground water studies (3),
aquatic monitoring (4), gene frequency mapping (5) and radionuclide
contamination from atmospheric nuclear tests (6-8).

Assumptions Used in Kriging

The early theoretical work was done by Matheron (9-11) at the Paris
School of Mines. Journel and Huijbregts (12), Rendu (13) and David
(14) have published books which describe the theoretical aspects of
kriging and the derivation of the kriging system of linear equations.

0097–6156/85/0292–0203$11.00/0
© 1985 American Chemical Society

Pauncz (15) and Bell and Reeves (16) have published bibliographies of
the English language publications on kriging. This section will
define some of the common terminology used in kriging and give an
overview of the assumptions made by kriging.

Random Functions and Regionalized Variables. In univariate
statistics, an observation y_i is defined as a realization of a random
variable Y, where Y has a probabilistic distribution (e.g., normal,
lognormal, exponential, etc.). This distribution is generally
characterized by certain parameters, such as the mean and variance,
which are assumed to exist but are unknown. Often the goal is to
make inferences about these unknown parameters. Consequently, a
number of realizations, say $\{y_1, \ldots, y_n\}$, of this random variable
are obtained and inferences about the parameters of the distribution
of the random variable are made using these observations. For
example, if we assume that Y has a normal distribution then the mean
(μ) and variance (σ^2) are estimated by

$$\hat{\mu} = \frac{1}{n} \sum_{i=1}^{n} y_i$$

$$\hat{\sigma}^2 = \frac{1}{n-1} \sum_{i=1}^{n} (y_i - \hat{\mu})^2$$

In kriging, the goal is to make inferences about some phenomena
which occurs over a continuous space. This phenomena is called a
random function, $Z(\underline{x})$, and is analogous to the random variable
described above. The regionalized variable (ReV) is a single
realization of the random function. In the univariate setting the
realization was a single observation, while in the spatial setting of
kriging, the ReV is a set of n observations and is denoted as

$$Z(\underline{x}_i) = \text{ReV observed at } \underline{x}_i \quad , \quad i = 1, \ldots, n$$

$$\underline{x}_i = (x_i, y_i), \text{ a location on a continuous space}$$
$$\text{(the space is two-dimensional in this case)}$$

An example of a random function is the distribution of lead in the
top 5 cm of soil within a mile radius of a lead smelter. An example
of a ReV would be the set of measurements obtained from taking say
100 core samples around the lead smelter. The important thing to
remember is that these 100 measurements constitute only a single
realization of the random function.

Second Order Stationarity. With only a single realization of the
random function it would be impossible to make any meaningful
inferences about the random function if we did not make some
assumptions about its stationarity. A random function is said to be
strictly stationary if the joint probability density function for k
arbitrary points is invariant under simultaneous translation of all

these points by any distance $\underline{h}$, that is

$$f(Z(\underline{x}_1), \ldots, Z(\underline{x}_k)) = f(Z(\underline{x}_1 + \underline{h}), \ldots, Z(\underline{x}_k + \underline{h}))$$

However, this assumption is not likely to hold for any realistic problem.

In practice, only the first two moments of the random function are of interest. The first order moment is the expectation (mean) of the random function at an arbitrary location $\underline{x}$, which is defined to be

$$E[Z(x)] = m(\underline{x})$$

The second order moment can be expressed either in terms of the covariance or the variogram. The covariance of the random function at points $\underline{x}_1$ and $\underline{x}_2$ is defined to be

$$COV[Z(\underline{x}_1), Z(\underline{x}_2)] = E[\{Z(\underline{x}_1) - m(\underline{x}_1)\}\{Z(\underline{x}_2) - m(\underline{x}_2)\}]$$

When $\underline{x}_1 = \underline{x}_2 = \underline{x}$, the covariance is simply the variance of the random function at $\underline{x}$, that is

$$VAR[Z(\underline{x})] = COV[Z(\underline{x}), Z(\underline{x})]$$

$$= E[\{Z(\underline{x}) - m(\underline{x})\}^2]$$

The semi-variogram is one-half the expected squared difference of an increment, $[Z(\underline{x}_1) - Z(\underline{x}_2)]$, that is

$$\gamma(\underline{x}_1; \underline{x}_2) = \frac{1}{2} E[(Z(\underline{x}_1) - Z(\underline{x}_2))^2] \qquad (1)$$

The above notation is often simplified by defining $\underline{h}$ to be the distance between locations $\underline{x}_1$ and $\underline{x}_2$. Thus, if $\underline{x}_2 = \underline{x}_1 + \underline{h}$, then

$$COV(\underline{h}) = COV[Z(\underline{x}_1), Z(\underline{x}_2)]$$

$$COV(\underline{0}) = VAR[Z(\underline{x})]$$

$$\gamma(\underline{h}) = \gamma(\underline{x}_1; \underline{x}_2)$$

The random function is said to be weakly or second order stationary when its first two moments are invariant under simultaneous translation by $\underline{h}$. That is, for every $\underline{x}$ and $\underline{h}$

$$E[Z(\underline{x})] = E[Z(\underline{x} + \underline{h})] = m(\underline{x}) = m < \infty$$

and $$COV(\underline{h}) = E[Z(\underline{x} + \underline{h}) \cdot Z(\underline{x})] - m^2 < \infty$$

That is, the random function has a constant mean and the covariance between each pair of locations depends only on the distance, $\underline{h}$, between the two observations. This also implies that

$$VAR[Z(\underline{x})] = COV(\underline{0}) < \infty$$

and $$2\gamma(\underline{h}) = VAR[Z(\underline{x}) - Z(\underline{x} + \underline{h})]$$
$$= COV(\underline{0}) - COV(\underline{h})$$

Thus, if the assumption of second order stationarity holds, then statistical inferences about the first two moments become possible since each pair of observations that are separated by a distance $\underline{h}$ can be considered a different realization of the random function.

<u>Intrinsic Hypothesis.</u> The assumption of second order stationarity assumes that the variance exists (i.e., it is not equal to infinity). This assumption is still stronger than necessary. A random function is said to be intrinsic (i.e., satisfies the intrinsic hypothesis) when for every $\underline{x}$

$$E[Z(\underline{x})] = m$$

and $$VAR[Z(\underline{x} + \underline{h}) - Z(\underline{x})] = 2\gamma(\underline{h})$$

for every $\underline{h}$. That is, only the increment $[Z(\underline{x} + \underline{h}) - Z(\underline{x})]$ has to have a variance and that variance does not depend on the location of $\underline{x}$.

To use what is termed simple kriging, only the assumption that the random function is intrinsic needs to be made. The problem with this assumption is that the expected value of the phenomena of interest is rarely a constant. For example, the expected concentration of lead in the soil around a smelter would decrease as the distance from the smelter increased. If this decrease (or trend) is gradual enough, it is often assumed that within a limited neighborhood the random function has a "local stationarity" and then simple kriging is used, since generally only the observations within the limited neighborhood are used in the estimation process.

<u>Intrinsic Random Function of Order K.</u> When the expected value of the random function cannot be assumed to be a constant, even within a limited neighborhood, then the random function is assumed to be the sum of two terms. That is,

$$Z(\underline{x}) = m(\underline{x}) + Y(\underline{x})$$

where $Y(\underline{x})$ is an intrinsic random function as described previously and $m(\underline{x})$ is a deterministic component, which is referred to as the drift. Then,

$$E[Z(\underline{x}) - Z(\underline{x} + \underline{h})] = m(\underline{x}) - m(\underline{x} + \underline{h})$$

$$VAR[Z(\underline{x}) - Z(\underline{x} + \underline{h})] = VAR[Y(\underline{x}) - Y(\underline{x} + \underline{h})] = 2\gamma(\underline{h})$$

but $$2\gamma(\underline{h}) = E[(Z(\underline{x}) - Z(\underline{x} + \underline{h}))^2] - (m(\underline{x}) - m(\underline{x} + \underline{h}))^2$$

Thus, to estimate the variogram, the drift must be known and to estimate the drift, the variogram must be known. This leads to difficulties in model identification which will be discussed later.

A way to avoid some of the difficulties mentioned above is to assume that $m(\underline{x})$ can be approximated, within a limited neighborhood, by a slowly varying polynomial of the form

$$m(\underline{x}) = \sum_{i=0}^{k} a_i f_i(\underline{x}) \qquad (2)$$

where a_i are constant coefficients and the f_i are monomials (e.g., if $\underline{x} = (x,y)$ then $f_0 = 1$, $f_1 = x$, $f_2 = y$, $f_3 = xy$, $f_4 = x^2$, $f_5 = y^2$, etc.) and use the concept of generalized incrementing (17). Generalized increments, which are analogous to the differencing process used in time series, "filter out" the drift. For example, the first order difference (zero order increment) filters out a constant. This is what happens in simple kriging where the drift is a constant, that is

$$Z(\underline{x}) = m + Y(\underline{x})$$

$$Z(\underline{x} + \underline{h}) - Z(\underline{x}) = m + Y(\underline{x} + \underline{h}) - m - Y(\underline{x}) = Y(\underline{x} + \underline{h}) - Y(\underline{x})$$

and the drift is "filtered out." When the drift is linear, then a second order difference (first order increment) will filter out the drift. Thus, if $\underline{x} = (x,y)$, then

$$Z(\underline{x}) = m(\underline{x}) + Y(\underline{x}) = a_0 + a_1 x + a_2 y + Y(\underline{x})$$

and $$Z(\underline{x} + \underline{h}) - 2Z(\underline{x}) + Z(\underline{x} - \underline{h}) = Y(\underline{x} + \underline{h}) - 2Y(\underline{x}) + Y(\underline{x} - \underline{h})$$

so that the first order generalized increment is now an intrinsic random function, and $Z(\underline{x})$ is termed an intrinsic random function of order 1 (IRF-1). In general, any sum

$$\sum_{i=1}^{n} \lambda_i Z(\underline{x}_i) \qquad \text{where } \underline{x}_i = (x_i, y_i)$$

for which $$\sum_{i=1}^{n} \lambda_i = 0 \ , \quad \sum_{i=1}^{n} \lambda_i x_i = 0 \ \text{ and } \ \sum_{i=1}^{n} \lambda_i y_i = 0$$

will filter out a linear drift and is a first order generalized increment. Thus the quantity

$$Z(-1, 0) + Z(1, 0) + Z(0, -1) + Z(0, 1) - 4\, Z(0, 0)$$

is also a generalized increment of order 1. A generalized increment of order 2 filters out a quadratic drift when

$$\sum_{i=1}^{n} \lambda_i = 0 \quad , \quad \sum_{i=1}^{n} \lambda_i x_i = 0 \quad , \quad \sum_{i=1}^{n} \lambda_i y_i = 0$$

$$\sum_{i=1}^{n} \lambda_i x_i y_i = 0 \quad , \quad \sum_{i=1}^{n} \lambda_i x_i^2 = 0 \quad \text{and} \quad \sum_{i=1}^{n} \lambda_i y_i^2 = 0$$

The second order increment is an intrinsic random function of order two (IRF-2).

To use what is termed universal kriging, it is assumed that $Z(\underline{x})$ is an intrinsic random function of order k. But the problem of identifying the drift and the semi-variogram when they are both unknown is still present. However, Matheron (11) defined a family of functions called the generalized covariance, $K(\underline{h})$, and the variance of the generalized increment of order k can be defined in terms of $K(\underline{h})$. That is,

$$\text{VAR}\left[\sum_{i=1}^{n} \lambda_i Z(\underline{x}_i)\right] = \sum_{i=1}^{n} \sum_{j=1}^{n} \lambda_i \lambda_j K(\underline{x}_i; \underline{x}_j)$$

The advantage of the generalized covariance is that its identification only requires that the order of the drift is known, not the values of the coefficients, a_i.

Simple kriging is actually a subset of universal kriging since the assumption that $Z(\underline{x})$ is an intrinsic random function of order 0 is the same as the assumption that $Z(\underline{x})$ is intrinsic. Additionally, when $Z(\underline{x})$ is intrinsic, the generalized covariance and the semi-variogram are related as follows:

$$\gamma(\underline{h}) = K(\underline{0}) - K(\underline{h}) \tag{3}$$

Kriging System of Linear Equations

In the previous section, an overview of the kriging assumptions was given. When these assumptions are accepted, a kriging system of linear equations can be developed. Whether the random function, $Z(\underline{x})$, is intrinsic (simple kriging) or it is an intrinsic random function of order k (universal kriging), and whether the semi-variogram or the generalized covariance is used, the kriging system of linear equations remains essentially the same. Additionally, whether kriging is used to estimate the value of $Z(\underline{x})$ at a point or over a given area, the kriging system of linear equations still remains essentially the same. Thus, in this section the kriging system of linear equations, using the semi-variogram, for estimating the average value of $Z(\underline{x})$ over a given area, when $Z(\underline{x})$ is an intrinsic random function of order 1 (i.e., $Z(\underline{x})$ has a linear drift), will be demonstrated. The modifications of this system for other situations will also be described.

Given a ReV which consists of n observations, $\{Z(\underline{x}_1), \ldots, Z(\underline{x}_n)\}$, from an intrinsic random function of order k, an estimate of a quantity Y which is any linear functional of $Z(\underline{x})$ is desired. The kriging estimator of Y is a weighted average of the data, i.e.,

$$\hat{Y} = \sum_{i=1}^{m} \lambda_i \, Z(\underline{x}_i) \tag{4}$$

where λ_i = the kriging weights

It should be noted that the weighted average in Equation 4 does not necessarily use all of the n observations (i.e., $m \leq n$). For example, when Y is the value of $Z(\underline{x})$ at a specific location $\underline{x}_0$, then generally only the m observations in the neighborhood of $\underline{x}_0$ are used in the weighted average. As the location of $\underline{x}_0$ changes, the m observations used in the weighted average also change, since the neighborhood has moved. Thus, when m < n, the kriging estimator of Y is a moving weighted average. The problem is to choose the weights in the best possible way.

The kriging system of linear equations is derived so that their solution gives kriging weights such that the kriging estimator is a "best linear unbiased estimator." The estimator is linear because the estimator is a weighted sum. It is unbiased because the system is constrained so that $E[\hat{Y} - Y] = 0$. It is "best" in the sense that within the class of all unbiased linear estimators, it has the smallest (minimum) mean square error. That is $E[(\hat{Y} - Y)^2]$ is a minimum. Since the estimator is unbiased the mean square error of the estimator is an estimate of the variance and is called the kriging variance, σ_k^2.

The most common quantities which are of interest are the value of $Z(\underline{x})$ at a specific location $\underline{x}_0$ (point estimation) and the average value of $Z(\underline{x})$ over a specified area (areal estimation). In the case of point estimation the quantity Y is simply defined as

$$Y = Z(\underline{x}_0)$$

In areal estimation the quantity Y is defined to be

$$Y = \frac{1}{A} \int_B Z(\underline{x})d\underline{x}$$

where B represents a block with an area A. The inventory or total amount of contaminant in the block can be estimated by using the total amount of contaminant in a core as the random function, then multiplying the areal estimate by the area within the block divided by the surface area of the core. In both cases, the kriging estimate is still a weighted sum, however the weights and the kriging variance will vary because the quantities used in the kriging system will be different.

The kriging system of linear equations is derived by minimizing the mean square error, $E[(\hat{Y} - Y)^2]$ under the unbiasedness constraint,

$E[\hat{Y} - Y] = 0$. It can be shown (see 13) that

$$E[(\hat{Y} - Y)^2] = -\sum_{i=1}^{m} \sum_{j=1}^{m} \lambda_i \lambda_j \gamma(\underline{x}_i;\underline{x}_j) - \overline{\gamma}(B;B) + 2 \sum_{j=1}^{m} \lambda_j \overline{\gamma}(\underline{x}_j;B)$$

where $\gamma(\underline{x}_i;\underline{x}_j)$ = value of the semi-variogram for the distance between $\underline{x}_i$ and $\underline{x}_j$

$$\overline{\gamma}(\underline{x}_j;B) = \frac{1}{A} \int_B \gamma(\underline{x}_j;\underline{x})d\underline{x}$$

= average semi-variogram between $\underline{x}_j$ and all the points in B

$$\overline{\gamma}(B;B) = \frac{1}{A^2} \int_B \int_B \gamma(\underline{x};\underline{x}')d\underline{x}d\underline{x}'$$

= average semi-variogram between any two points $\underline{x}$ and $\underline{x}'$ sweeping independently throughout B

For point estimation, $B = \underline{x}_0$, thus $\overline{\gamma}(\underline{x}_j;B) = \gamma(\underline{x}_j;\underline{x}_0)$ and $\overline{\gamma}(B;B) = \overline{\gamma}(\underline{x}_0;\underline{x}_0) = 0$ (by the definition of the semi-variogram). The unbiasedness constraint, $E[\hat{Y} - Y] = 0$, is satisfied when

$$\sum_{i=1}^{m} \lambda_i f_t(\underline{x}_i) = \overline{f}_t(B) \qquad \text{for } t = 0, ..., s$$

where $\overline{f}_t(B) = \frac{1}{A} \int_B f_t(\underline{x})d\underline{x}$

and the $f_t(\underline{x})$ are the drift monomials in Equation 2. For point estimation, $\overline{f}_t(B) = f_t(\underline{x}_0)$, thus when $\underline{x}_0 = (x_0, y_0)$, $f_0(\underline{x}) = 1$, $f_1(\underline{x}_0) = x_0$, $f_2(\underline{x}_0) = y_0$, $f_3(\underline{x}_0) = x_0 y_0$, etc. When the drift is constant (simple kriging), $s = 0$, $f_0(\underline{x}_i) = \overline{f}_0(B) = 1$ and the constraint is

$$\sum_{i=1}^{m} \lambda_i = 1$$

When the drift is linear ($Z(\underline{x})$ is an IRF-1), then $s = 2$, and

$$\overline{f}_1(B) = \frac{1}{A} \int_B x dx dy = \overline{x}$$

$$\overline{f}_2(B) = \frac{1}{A} \int_B y \, dy \, dx = \overline{y}$$

and the constraints are

$$\sum_{i=1}^{m} \lambda_i = 1 \quad , \quad \sum_{i=1}^{m} \lambda_i x_i = \overline{x} \quad \text{and} \quad \sum_{i=1}^{m} \lambda_i y_i = \overline{y}$$

The minimization of $E[(\hat{Y} - Y)^2)]$ with the constraint $E[\hat{Y} - Y] = 0$ is done using $s+1$ Lagrange multipliers $(\mu_0, \ldots, \mu_s)$. The result is the following kriging system of linear equations

$$\sum_{i=1}^{m} \lambda_j \gamma(\underline{x}_i; \underline{x}_j) + \sum_{t=0}^{s} \mu_t f_t(\underline{x}_i) = \overline{\gamma}(\underline{x}_i; B) \qquad \text{for} \quad j = 1, \ldots, m$$

(5)

$$\sum_{i=1}^{m} \lambda_i f_t(\underline{x}_i) = \overline{f}_t(B) \qquad \text{for} \quad t = 0, \ldots, s$$

The kriging variance is

$$\sigma_k^2 = \sum_{i=1}^{m} \lambda_i \overline{\gamma}(\underline{x}_i; B) + \sum_{t=0}^{s} \mu_t \overline{f}_t(B) - \overline{\gamma}(B; B) \qquad (6)$$

The solution of this system for the kriging weights, λ_i, is best done using matrix algebra. When $Z(\underline{x})$ is an IRF-1 define

$$\underline{A} = \begin{bmatrix} \gamma(\underline{x}_1; \underline{x}_1) & \gamma(\underline{x}_1; \underline{x}_2) & \cdots & \gamma(\underline{x}_1; \underline{x}_m) & 1 & x_1 & y_1 \\ \gamma(\underline{x}_2; \underline{x}_1) & \gamma(\underline{x}_2; \underline{x}_2) & \cdots & \gamma(\underline{x}_2; \underline{x}_m) & 1 & x_2 & y_2 \\ \vdots & \vdots & & \vdots & \vdots & \vdots & \vdots \\ \gamma(\underline{x}_m; \underline{x}_1) & \gamma(\underline{x}_m; \underline{x}_2) & \cdots & \gamma(\underline{x}_m; \underline{x}_m) & 1 & x_m & y_m \\ 1 & 1 & \cdots & 1 & 0 & 0 & 0 \\ x_1 & x_2 & \cdots & x_m & 0 & 0 & 0 \\ y_1 & y_2 & \cdots & y_m & 0 & 0 & 0 \end{bmatrix}$$

$$\underline{B}' = [\lambda_1, \lambda_2, \ldots, \lambda_m, \mu_0, \mu_1, \mu_2]$$

$$\underline{C}' = [\overline{\gamma}(\underline{x}_1; B), \overline{\gamma}(\underline{x}_2; B), \ldots, \overline{\gamma}(x_m; B), 1, \overline{x}, \overline{y}]$$

$$\underline{D}' = [Z(\underline{x}_1), Z(\underline{x}_2), \ldots, Z(\underline{x}_m), 0, 0, 0]$$

then the kriging system is

$$\underline{A}\ \underline{B} = \underline{C}$$

and the solution of this system for the kriging weights and Lagrange multipliers is

$$\underline{B} = \underline{A}^{-1}\underline{C}$$

Therefore, the kriging estimator and variance are

$$\hat{Y} = \underline{B}'\ D$$

$$\sigma_k^2 = \underline{B}'\underline{C} - \overline{\gamma}(B;B)$$

When the generalized covariance is used instead of the semi-variogram, the kriging system looks almost the same as the one above because of the relationship between the generalized covariance and semi-variogram shown in Equation 3. Thus, the semi-variogram $\gamma(\underline{x}_i;\underline{x}_j)$ is replaced by $K(\underline{x}_i;\underline{x}_j)$ in matrix A and $\overline{\gamma}(\underline{x}_i;B)$ is replaced by $\overline{K}(\underline{x}_i;B)$ in vector $\underline{C}'$, and the kriging variance is now

$$\sigma_k^2 = \overline{K}(B;B) - \underline{B}'\underline{C}$$

Model Estimation

As seen in Equation 4, the kriging estimator of the value of $Z(\underline{x})$ at a specific location or the average value of $Z(\underline{x})$ over a specified area is a weighted average of the data. The kriging weights used in the weighted average and the kriging variance are obtained from solving the kriging system of linear equations as shown in Equation 5. When the model (i.e., drift and semi-variogram or generalized covariance) is known, the kriging estimator is a best linear unbiased estimator (BLUE). However, the model is generally unknown and thus must be estimated using the observations. If the model is not identified correctly, the kriging estimator is no longer BLUE. In this section the estimation of the model is described.

Semi-variogram Models. The semi-variogram is a function of distance (h). That is, the semi-variogram at h is one half the expected squared difference between a pair of observations $Z(\underline{x})$ that are separated by a distance h (see Equation 1). This function (or model) must be conditionally positive definite so that the variance of the linear functional of $Z(\underline{x})$ is greater than or equal to zero. Five of the common semi-variogram models which satisfy this condition are:

1. Power Function (Figure 1a)

$$\gamma(h) = b|h|^p \quad \text{with } 0 < p < 2$$

(when p = 1, the semi-variogram is a linear model)

2. Spherical Model (Figure 1b)

$$\gamma(h) = C[\frac{3}{2} \frac{|h|}{a} - \frac{1}{2} \frac{|h|^3}{a^3}] \quad \text{for } |h| \leq a$$

$$\gamma(h) = C \qquad\qquad\qquad \text{for } |h| > a$$

'a' equals the range of the semi-variogram and C equals the sill. The range can be thought of as the "zone of influence." If the distance between two points is less than the range, then the value at one point is correlated with the value at the other point. If the distance between two points is greater than the range, then the points are independent. The sill is the bound on the semi-variogram and provides an estimate of the overall variability. When a semi-variogram is bounded then the random function is second order stationary and

$$COV[Z(\underline{x} + \underline{h}), Z(\underline{x})] = VAR[Z(\underline{x})] - \gamma(\underline{h})$$

When $|\underline{h}| > a$

$$COV[Z(\underline{x} + \underline{h}), Z(\underline{x})] = 0$$

$$\gamma(\underline{h}) = C$$

and thus

$$VAR[Z(\underline{x})] = C$$

3. Cubic Model (Figure 1c)

$$\gamma(h) = C[7 \frac{|h|^2}{a^2} - \frac{35}{4} \frac{|h|^3}{a^3} + \frac{7}{2} \frac{|h|^5}{a^5} - \frac{3}{4} \frac{|h|^7}{a^7}] \quad \text{for } |h| \leq a$$

$$\gamma(h) = C \qquad \text{for } |h| > a$$

4. Exponential Model (Figure 1d)

$$\gamma(h) = C[1 - \bar{e}^{|h|/a}]$$

(the range of the semi-variogram is approximately 3a)

5. Gaussian Model (Figure 1e)

$$\gamma(h) = C[1 - e^{-(h/a)^2}]$$

(the range of the semi-variogram is approximately 2a)

When h is set to zero, $\gamma(0)$ must also be equal to zero. However, if the difference $[Z(\underline{x}) - Z(\underline{x}')]$ does not tend to zero, for measurements taken at arbitrarily close points $\underline{x}$ and $\underline{x}'$, then there

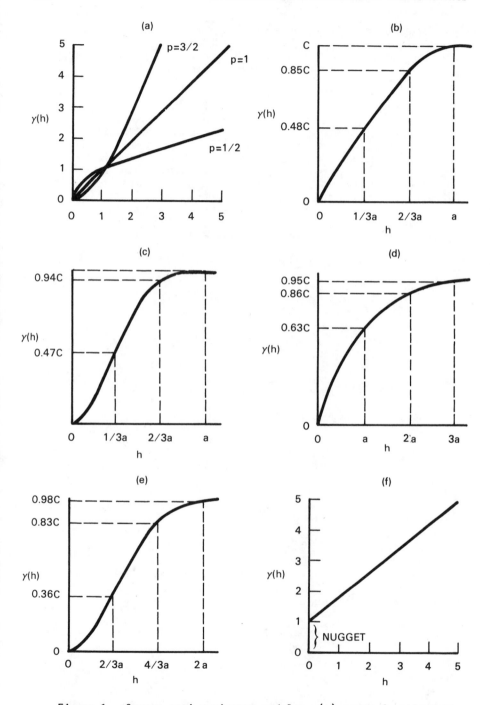

Figure 1. Common semi-variogram models: (a) power function, (b) spherical, (c) cubic, (d) exponential, (e) Gaussian and (f) linear with a nugget.

is a discontinuity of the semi-variogram at the origin. This
discontinuity is called the nugget effect. If there is a nugget
effect, the semi-variogram model is adjusted to take it into account.
For example if the model is linear with a nugget of size K
(Figure 1f) then

$$\gamma(h) = b|h| + K \qquad \text{for} \qquad |h| > 0$$

$$\gamma(h) = 0 \qquad \text{for} \qquad |h| = 0$$

Estimation of the Semi-variogram When the Drift is Constant. In
practice, the semi-variogram at each distance h is estimated as
follows:

$$\hat{\gamma}(h) = \frac{1}{2N(h)} \sum_{i=1}^{N(h)} [Z(\underline{x}_i + \underline{h}) - Z(\underline{x}_i)]^2 \qquad (7)$$

where N(h) is the number of pairs of points, a distance h apart,
actually taken into the sum. Then one of the above common models is
fitted to the estimates of the semi-variogram at each h. Usually
$\hat{\gamma}(h)$ for each h is based on a range of distances, since insufficient
data exists for a specific distance h.

Once $\gamma(h)$ has been estimated, the correct semi-variogram model,
usually one of the five models discussed above, has to be selected
and the parameters of the model need to be estimated. Most model
fitting is done by "trial and error." Generally the appropriate
model can be chosen visually. For example, if the variogram has a
sigmoid shape, then either the cubic or Gaussian model is
appropriate. To distinguish between these two models, note that the
relationship between the sill and range are different: for the cubic
model, $\gamma(1/3a) = 0.47C$, while for the Gaussian model, $\gamma(1/3a) =$
0.36C. The estimation of the parameters (the sill and range) for a
given semi-variogram model is again governed by the relationship
between these parameters. Once the model is chosen and the sill is
estimated, then the range is set. The estimate of the sill and range
can be adjusted to some extent to improve the "fit" of the model.
However, it should be noted that "small" changes in the parameters of
the semi-variogram model do not make a significant difference in the
kriging weights and variance which are calculated by solving the
kriging system of linear equations. Thus, this procedure seems to be
no worse than any other technique.

Estimation of the Semi-variogram When the Drift is Not Constant.
When a non-constant drift is present, the estimation of the
semi-variogram model is confounded with the estimation of the drift.
That is, to find the optimal estimator of the semi-variogram, it is
necessary to know the drift function, but it is unknown. David (14)
recommended an estimator of the drift, m*(x), derived from
least-square methods of trend surface analysis (18). Then at every
data point a residual is given by

$$Y^*(\underline{x}) = Z(\underline{x}) - m^*(\underline{x})$$

An experimental variogram of estimated residuals $\gamma^*(h)$ can then be calculated. However, this variogram differs from the underlying variogram of the true residuals, $\gamma(h)$, and the bias is a function of the form of the estimator $m^*(\underline{x})$. In order to find $\gamma(h)$ from $\gamma^*(h)$, $\gamma^*(h)$ is graphically compared with a set of $\gamma_0^*(h)$ defined from various types of variograms $\gamma_0(h)$ and the same type of estimator $m^*(\underline{x})$. If the fit is "reasonable" (there is no test for the goodness of fit), the model $\gamma(h)$ is assumed correct. If the fit is not "reasonable" the process starts again with a new estimator of the drift.

Neuman and Jacobson (19) have developed a step-wise iterative regression process for simultaneously estimating the global drift and residual semi-variogram. Estimates of the function are obtained by solving a modified set of simple kriging equations written for the residuals. The modifications consist of replacing the true semi-variogram in the kriging equations by the semi-variogram of the residual estimates as obtained from the iterative regression process.

Generalized Covariance Models. When $Z(\underline{x})$ is an intrinsic random function of order k, an alternative to the semi-variogram is the generalized covariance (GC) function of order k. Like the semi-variogram model, the GC model must be a conditionally positive definite function so that the variance of the linear functional of $Z(\underline{x})$ is greater than or equal to zero. The family of polynomial GC functions satisfy this requirement. The polynomial GC of order k is

$$K(h) = C\delta - \sum_{i=0}^{k} (-1)^i \alpha_i |h|^{2i+1}$$

where C is the nugget effect which was described earlier and

$$\delta = \begin{array}{lll} 1 & \text{if} & h = 0 \\ 0 & \text{if} & h \neq 0 \end{array}$$

When $k \leq 2$ and $\underline{x}$ is two-dimensional, the coefficients α_i have the following constraints: $\alpha_0 \geq 0$, $\alpha_2 \geq 0$ and

$$\alpha_1 \geq \frac{-10}{3} \sqrt{\alpha_0 \alpha_2}$$

The order of the polynomial GC model is the same as the order of the drift. Thus the available models can be summarized as follows:

DRIFT	k	POLYNOMIAL GC MODEL
Constant	0	$C\delta - \alpha_0|h|$
Linear	1	$C\delta - \alpha_0|h| + \alpha_1|h|^3$
Quadratic	2	$C\delta - \alpha_0|h| + \alpha_1|h|^3 - \alpha_2|h|^5$

As can be seen above, when the drift is constant, the GC models are quite limited (i.e., $K(h) = C\delta$, $K(h) = -\alpha_0|h|$ or $K(h) = C\delta - \alpha_0|h|$). Thus, when there is a constant drift, the semi-variogram models should be used instead of GC models.

Estimation of the Generalized Covariance Model. An algorithm for the estimation of the order of the drift and the coefficient of the polynomial GC function has been developed by Delfiner (17). This algorithm, termed "Automatic Structure Identification (ASI)" is used in BLUEPACK 3D (a proprietary computer package sold by the Paris School of Mines) and is only briefly described in the literature (17). A similar algorithm has been developed by Hughes and Lettenmeier (20), who included the computer programs in their publication.

The ASI algorithm is broken down into three steps. First the order of the drift (k) is estimated. Then all the possible polynomial GC models are estimated. Thus if k = 0, there are 3 models estimated; k = 1, there are 7 models estimated and if k = 2, there are 15 models estimated. The inadmissible models (those models whose parameter estimates do not meet the constraints of the polynomial GC model) are discarded and the three best models are chosen. The third step compares the remaining models and makes the final choice.

The ASI method has the advantage of being automated. However, this method has its problems: the order of the drift tends to be underestimated when samples are from a symmetric grid (symmetric neighborhoods tend to filter polynomials by itself); the final choice of the model depends on an ad-hoc decision procedure (there are again no goodness of fit tests); and when the drift is constant, the only model is linear with a nugget which is not a large enough class of models, thus the user needs to go back to the variogram analysis described earlier. Probably its biggest weakness is its lack of robustness against variables that do not well satisfy the intrinsic hypothesis.

Kriging Analysis of Lead Measurements in Soil Cores

The lead measurement ($Z(\underline{x})$ = ppm lead in a soil core) are from three sites; RSR and DMC are centered around lead smelters while REF is a reference or control site. RSR has 208 measurements, DMC has 206 measurements and REF has 100 measurements. Figures 2 through 4 display the spatial distribution of the measurements at RSR, DMC and REF, respectively. The distribution of the data for all three sites are skewed to the right. Therefore, the natural logarithm (LN) of the data is used in the kriging analysis.

It is assumed that the intrinsic hypothesis holds within the limited neighborhood that is used in calculating the kriging estimates. That is, eight measurements will be used by the kriging estimator and within the limited neighborhood (with a radius of approximately 1000 feet) that these measurements occur it is assumed that there is no drift or systematic trend.

Semi-variogram Estimation. The first step in the kriging analysis is to estimate the semi-variogram for each site. The sample

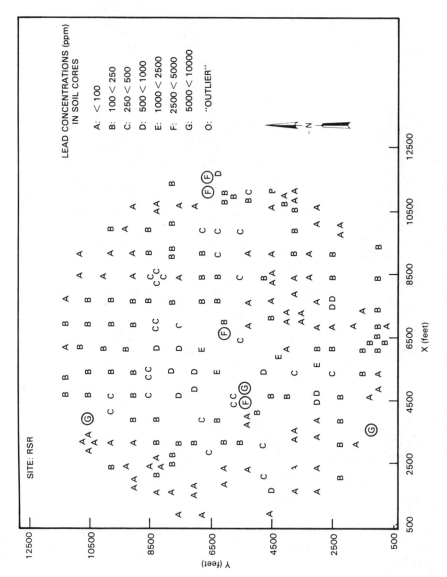

Figure 2. Display of measurements at RSR.

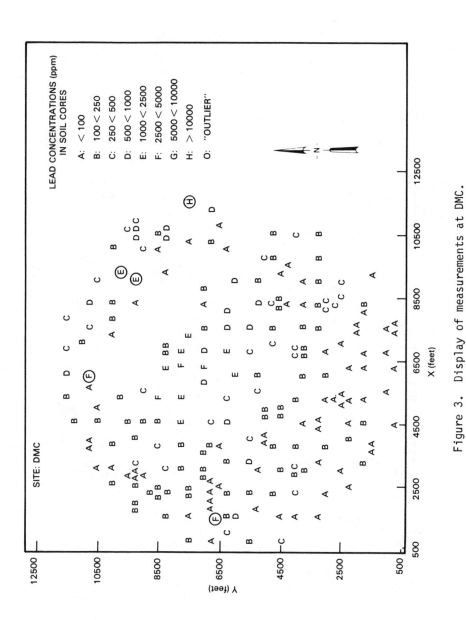

Figure 3. Display of measurements at DMC.

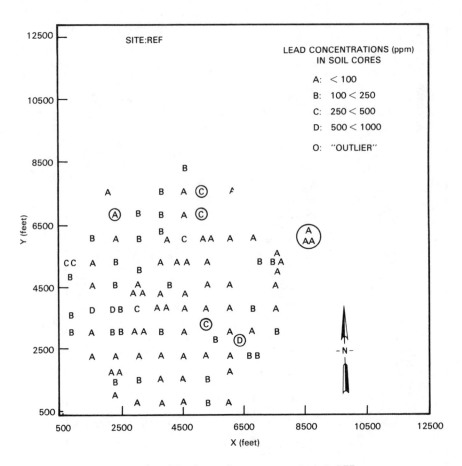

Figure 4. Display of measurements at REF.

semi-variogram is very sensitive to "outliers," thus to obtain a good estimate of the underlying semi-variogram, the data needs to be scrutinized and any "outliers" removed.

It can be seen in Figures 2 through 4 that some of the "outliers" are easy to spot and generally the justification for removing them is evident. For example, the measurements over 10,000 ppm at DMC was taken near a junk yard. It can be argued that the lead in this sample was not primarily from the DMC source, but a result of a source within the junk yard. Additionally, three measurements at REF are basically outside the range of the sampling pattern. Since these three points are all within approximately a foot of each other and would be considered as only one measurement in the kriging analysis, they would have little impact on the kriging estimates. In the end, 7 "outliers" are removed from RSR, 5 "outliers" are removed from DMC and 8 "outliers" are removed from REF.

Figures 5 through 7 show the sample semi-variograms for each site after the "outliers" are removed. When looking at the sample semi-variograms for RSR and DMC, they appear to be different realizations of the same underlying phenomena. Both sites are in the same general area (city) and the variable (lead concentration) is dispersed by the same process (a smelter). Therefore, to get a better estimate of the semi-variogram, the semi-variograms for DMC and RSR are combined (see Figure 8).

Among the common semi-variogram models, the exponential model best fits the sample semi-variograms. Therefore, for REF

$$\hat{\gamma}(h) = 0.35[1 - e^{h/200}] + 0.01$$

is used. For DMC and RSR

$$\hat{\gamma}(h) = 1.40[1 - e^{h/2200}] + 0.01$$

is used.

Kriging Estimates and Standard Deviations. The kriging analysis is performed on the natural logarithms of the measurements, the kriging estimator $\hat{Y}_i$, is in log scale. $EXP[\hat{Y}_i]$ is not an unbiased estimate of the mean concentration in the block, it is an estimate of the median block value. Rendu (21) shows that the unbiased kriging estimator of the mean concentration in the original scale is

$$Y^* = EXP[\hat{Y}_L + \sum\lambda_i\overline{\gamma}(x_i;B) - \frac{1}{2}\{\sigma_{kL}^2 + \overline{\gamma}(B;B)\}]$$

where λ_i are the kriging weights and σ_{kL}^2 is the logarithmic kriging variance.

The kriging estimates of the mean concentration (ppm lead) over a 250 foot by 250 foot block and the kriging standard deviation for each block are shown in Figures 9 through 14. At RSR and DMC the estimated block means are shown for blocks whose multiplicative kriging standard deviation was less than 2. (Since the measurements are transformed using the natural logarithm, the standard deviations

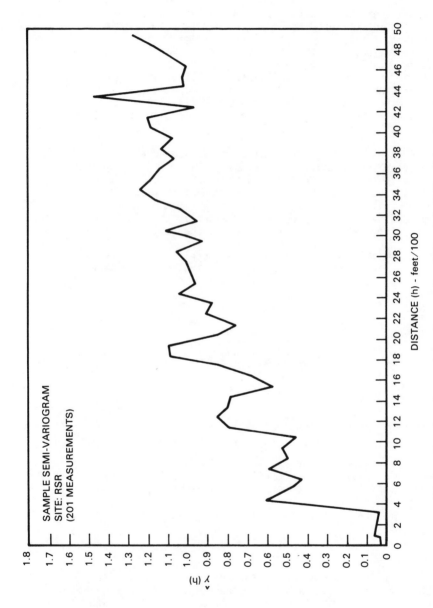

Figure 5. Sample semi-variogram for RSR with 7 "outliers" removed.

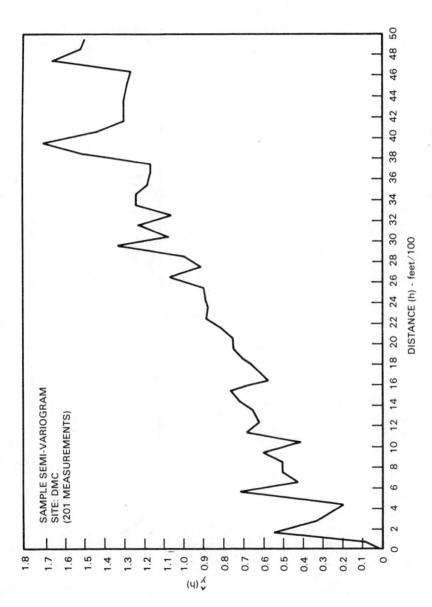

Figure 6.   Sample semi-variogram for DMC with 5 "outliers" removed.

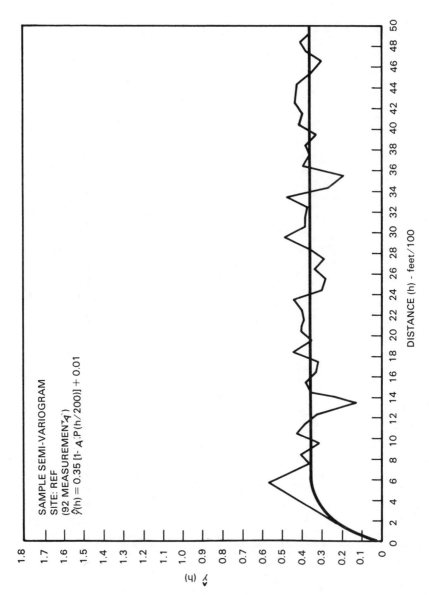

SAMPLE SEMI-VARIOGRAM
SITE: REF
(92 MEASUREMENTS)
$\hat{\gamma}(h) = 0.35\,[1 - A \cdot P\,(h/200)] + 0.01$

$\hat{\gamma}$ (h)

DISTANCE (h) - feet/100

Figure 7.   Sample semi-variogram for REF with 8 "outliers"
removed.

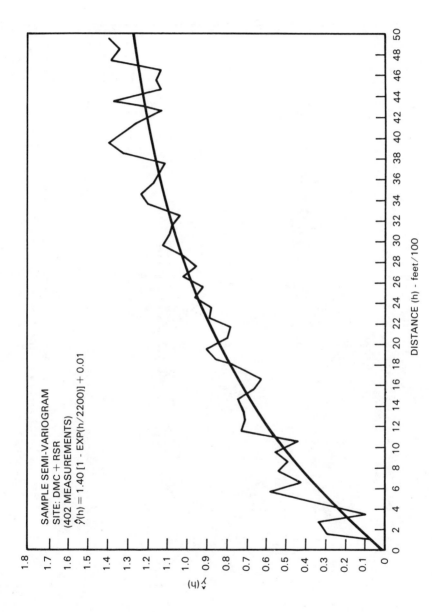

Figure 8.   Combined sample semi-variogram for DMC and RSR.

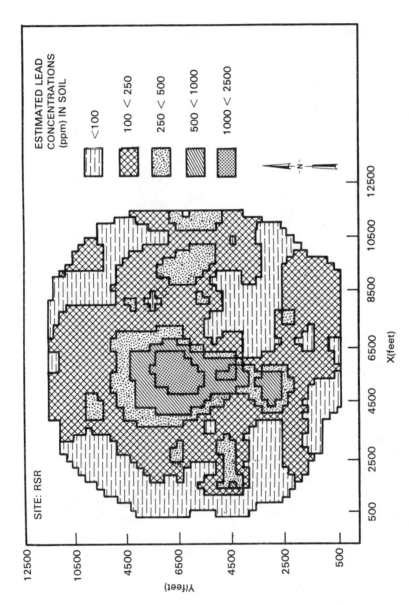

Figure 9. Kriging estimates of the mean lead concentration for 250 ft by 250 ft blocks at RSR.

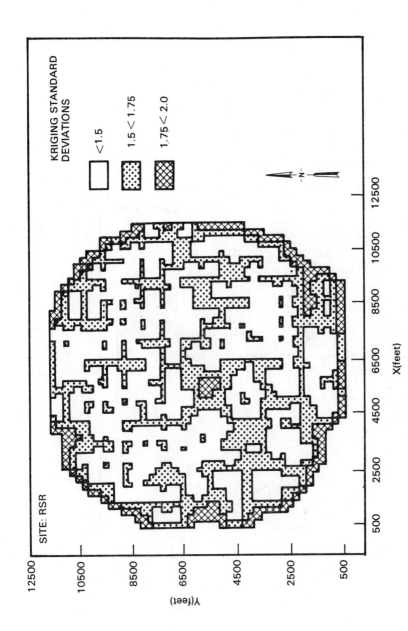

Figure 10. Multiplicative kriging standard deviations of the mean lead concentration over 250 ft by 250 ft blocks at RSR.

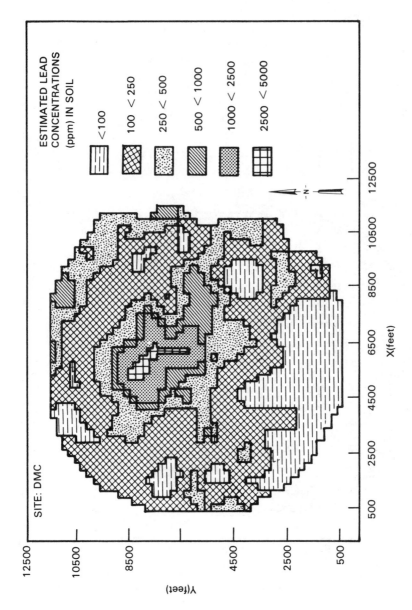

Figure 11. Kriging estimates of the mean lead concentration for 250 ft by 250 ft blocks at DMC.

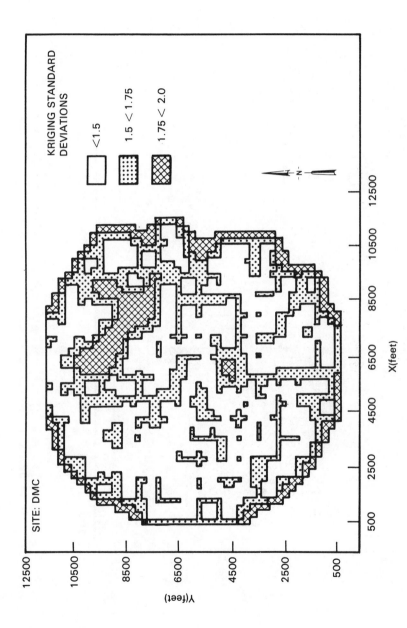

Figure 12. Multiplicative kriging standard deviations of the mean lead concentration over 250 ft by 250 ft blocks at DMC.

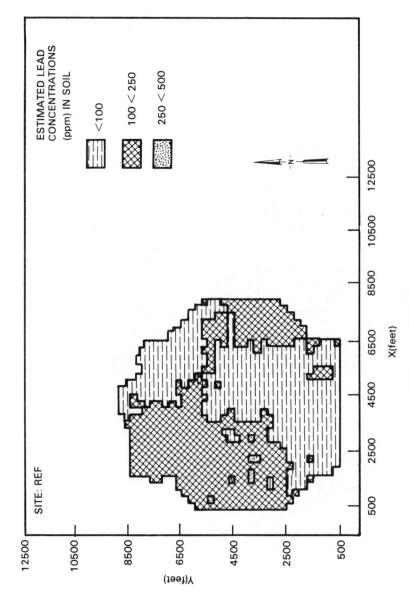

Figure 13. Kriging estimates of the mean lead concentration for 250 ft by 250 ft blocks at REF.

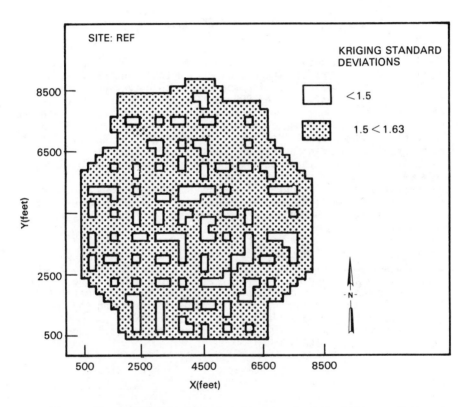

Figure 14. Multiplicative kriging standard deviations of the mean lead concentration over 250 ft by 250 ft blocks at REF.

are multiplicative when transformed back into the original scale.)
At REF the estimated block means are shown for blocks whose
multiplicative kriging standard deviation is less than 1.63. The
blocks that are not shown were outside the area that was sampled.

Confidence Intervals (or Bands). The 80% confidence interval about
the true mean for each individual block is calculated. Since the
kriging is done on the natural logarithm, the kriging standard
deviation is multiplicative and the 80% confidence interval is
approximately

$$Y*/EXP[1.2816 \; \sigma_{kL}] < \mu < Y*EXP[1.2816 \; \sigma_{kL}]$$

The 80% confidence bands for a given concentration level are
constructed such that all the blocks within the band are those whose
80% confidence interval contains the given concentration level. That
is, if we want to estimate the 80% confidence band for 250 ppm lead,
all those blocks whose lower limits are greater than 250 ppm lead are
classified as blocks whose concentrations are greater than 250 ppm.
Those blocks whose upper limits are less than 250 ppm are classified
as blocks whose concentrations are less than 250 ppm. The blocks
which are left over, those containing 250 ppm in the 80% confidence
interval, constitute the confidence band about the 250 ppm
concentration level. Figures 15 through 22 show the 80% confidence
bands for 2500 ppm, 1000 ppm, and 500 ppm concentration levels for
the RSR and DMC and 500 ppm and 250 ppm concentration levels for REF,
respectively.

Interpretation of Kriging Results. As seen in Figures 9 through 22,
the interpretation of the kriging results is primarily visual. These
figures allow the viewer to quickly access the extent of the
estimated contamination and the variability in those estimates.
Additionally, these figures provide a means of assessing the
confidence level of the estimated lead contamination.
     The sampled area (or the area in which kriging estimates are
retained) for each site are

Site	Number of Blocks	Area (Acres)
RSR	1717	2463.556
DMC	1676	2403.294
REF	834	1196.625

where each block was 250 ft by 250 ft (1.435 acres). It can be seen
in Figure 13 that 99.28% of the area at the control site (REF) has
estimated lead concentrations of less than 250 ppm and 100% of the
area has estimated lead concentrations of less than 500 ppm.
Additionally, from Figure 21 it can be seen that, at the 80%
confidence level, 99.64% of the area has estimated lead concentrations of less than 500 ppm. Therefore, with some assurance, it can
be stated that the "background level" of lead concentrations is no
greater than 500 ppm.

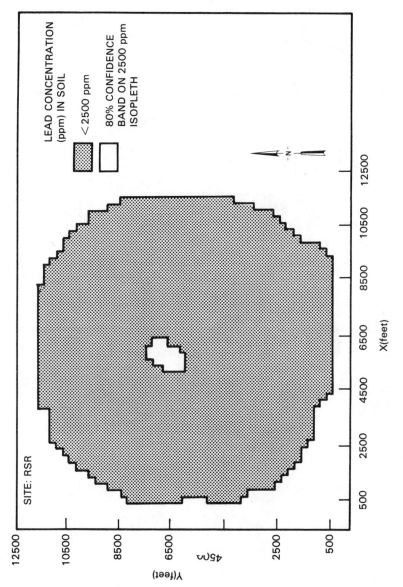

Figure 15. 80% confidence band for 2500 ppm lead isopleth at RSR.

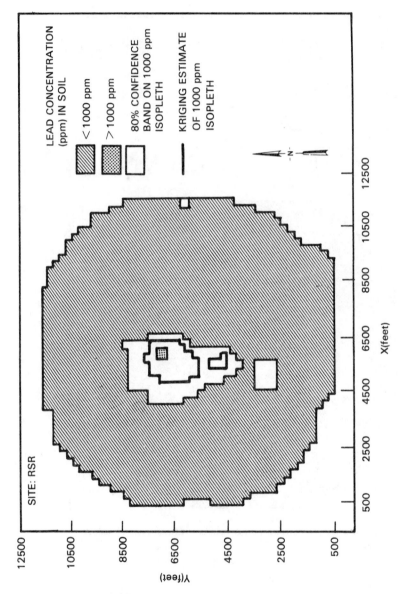

Figure 16. 80% confidence band for 1000 ppm lead isopleth at RSR.

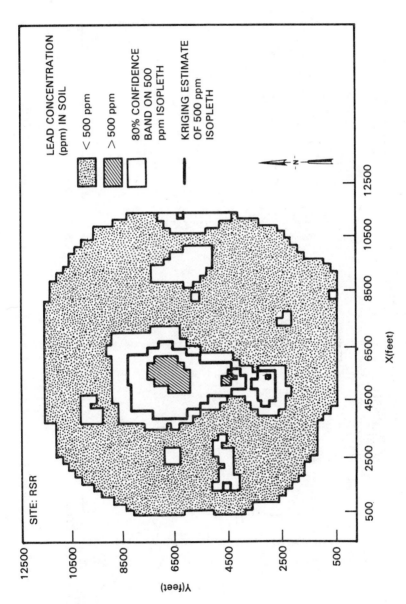

Figure 17. 80% confidence band for 500 ppm lead isopleth at RSR.

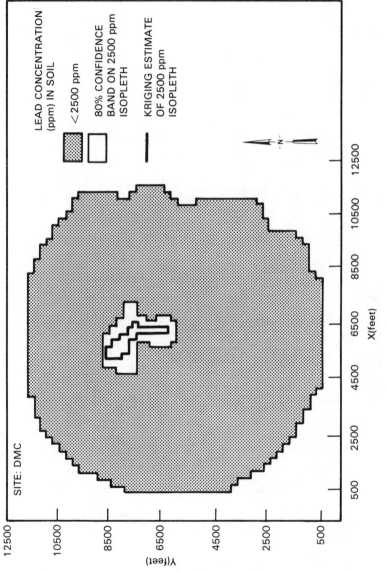

Figure 18. 80% confidence band for 2500 ppm lead isopleth at DMC.

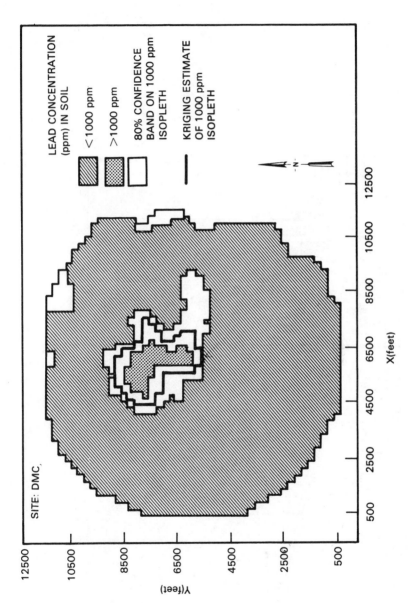

Figure 19. 80% confidence band for 1000 ppm lead isopleth at DMC.

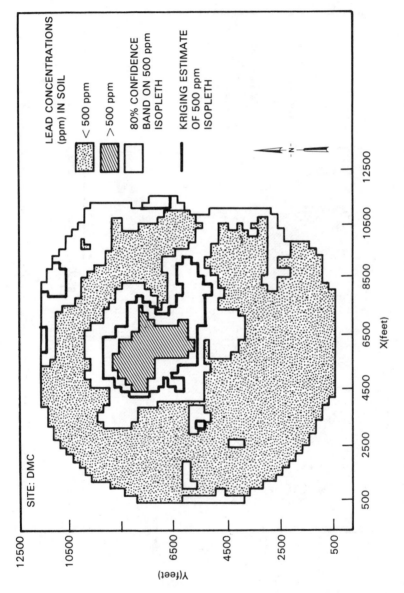

Figure 20. 80% confidence band for 500 ppm lead isopleth at DMC.

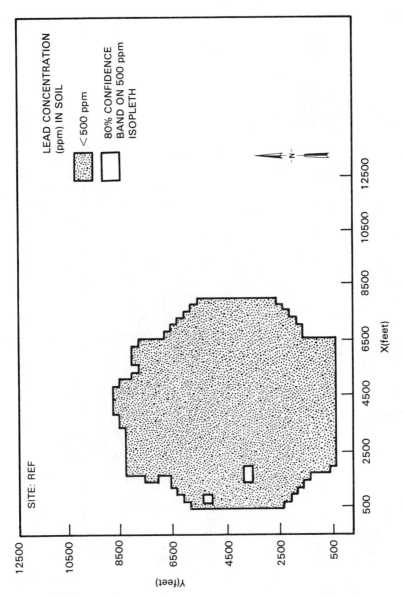

Figure 21. 80% confidence band for 500 ppm lead isopleth at REF.

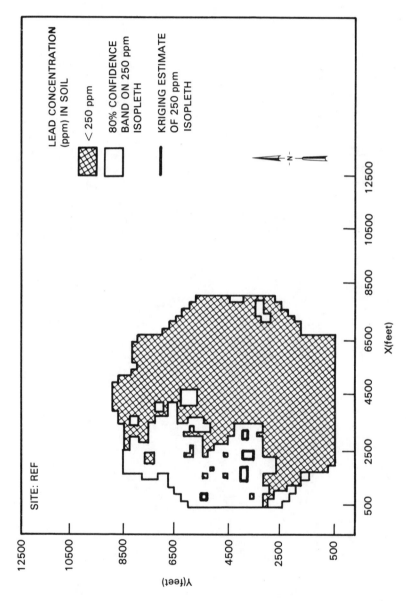

Figure 22. 80% confidence band for 250 ppm lead isopleth at REF.

It can be seen in Figure 9 that 8.39% of the area (almost 207 acres) at RSR is estimated to have lead concentrations above the "background level." Additionally, from Figure 17 it can be seen that at the 80% confidence level at least 2.39% of the area (almost 59 acres) is above the "background level" and there could be over 20.09% of the area (almost 495 acres) above the "background level."

It can be seen in Figure 11 that 11.52% of the area (almost 277 acres) at DMC is estimated to be above the "background level." Additionally, from Figure 20 it can be seen that at the 80% confidence level at least 5.25% of the area (over 126 acres) is above the "background level" and there could be over 33.37% of the area (almost 802 acres) above the "background level."

Acknowledgments

This research is sponsored by the U.S. Environmental Protection Agency, Office of Research and Development, Environmental Monitoring Systems Laboratory, under a related services agreement with the U.S. Department of Energy, Contract DE-AC06-76RLO 1830.

Literature Cited

1. Chiles, J. P. In "Advanced Geostatistics in the Mining Industry"; Guarascio, M.; et al., Eds.; D. Reidel Publishing Co.: Dordrecht-Holland, 1976; pp. 69-90.
2. Doctor, P. G.; Nelson, R. W. "Geostatistical Estimation of Parameters for Transport Modeling"; Pacific Northwest Laboratory: Richland, Washington, 1980; PNL-SA-8482.
3. Delhomme, J. P. Adv. Water Resour. 1978, 1, 251-6.
4. Hughes, J. P.; Lettenmeier, D. P. Water Resour. Res. 1981, 17, 1641-50.
5. Piazza, A.; Menozzi, P.; Cavall-Sforza, L. "The Making and Testing of Geographic Gene Frequency Maps"; Dept. of Genetics, School of Medicine, Stanford University: Stanford, California, 1979.
6. Barnes, M. G. "Statistical Design and Analysis in the Cleanup of Environmental Radionuclide Contamination"; Desert Research Institute, University of Nevada: Las Vegas, NV, 1978; NVO 1253-12.
7. Delfiner, P.; Gilbert, R. O. In "Selected Environmental Plutonium Research Reports of the NAEG"; White, M. G.; Dunaway, P. B., Eds.; Department of Energy: Las Vegas, NV, 1978; NVO-192, Vol. 2, pp. 405-450.
8. Simpson, J. C.; Gilbert, R. O. "Estimates of ^{239} ^{240}Pu + ^{241}Am Inventory, Spatial Pattern and Soil Tonnage for Removal at Nuclear Site-201, NTS"; Pacific Northwest Laboratory: Richland, Washington, 1980; PNL-SA-8269.
9. Matheron, G. Economic Geology 1963, 58, 1246-66.
10. Matheron, G. "The Theory of Regionalized Variables and Its Applications"; Ecole des Mines de Paris: Fontainebleau, France, 1971; No. 5.
11. Matheron G. Adv. Appl. Prob. 1973, 5, 439-68.
12. Journel, A. G.; Huijbregts, C. J. "Mining Geostatistics"; Academic Press: London, 1978.

13. Rendu, J. M. "An Introduction to Geostatistical Methods of Mineral Evaluation"; South African Institute of Mining and Metallurgy: Johannesburg, 1978.
14. David, M. "Developments in Geomathematics 2, Geostatistical Ore Reserve Estimation"; Elsevier Scientific Publishing Company: New York, 1977.
15. Pauncz, I. Math. Geology 1978, 10, 253-60.
16. Bell, G. D.; Reeves, M. Proc. Australas. Inst. Min. Metall. 1979, 169, 17.
17. Delfiner, P. In "Advanced Geostatistics in the Mining Industry"; Guarascio, M.; et al., Eds.; D. Reidel Publishing Co.: Dordrecht-Holland, 1976; pp. 49-68.
18. Agterberg, F. P. "Geomathematics"; Elsevier: The Netherlands, 1974.
19. Neuman, S. P.; Jacobson, E. A. Math Geology 1984, 16, 499-521.
20. Hughes, J. P.; Lettenmeier, D. P. "Aquatic Monitoring: Data Analysis and Network Design Using Regionalized Variable Theory"; Charles W. Harris Hydraulics Laboratory, University of Washington: Seattle, Washington, 1980; No. 65.
21. Rendu, J. M. In "Application of Computers and Operations, Research in the Mining Industry"; O'Neil, T. J., Ed.; Society of Mining Engineers of AIME, University of Arizona: Tempe, Arizona, 1979; pp. 199-212.

RECEIVED June 28, 1985

Simple Modeling by Chemical Analogy Pattern Recognition

W. J. Dunn III[1], Svante Wold[2], and D. L. Stalling[3]

[1] Department of Medicinal Chemistry and Pharmacognosy, University of Illinois at Chicago, Chicago, IL 60612
[2] Research Group for Chemometrics, Umea University, S901 87 Umea, Sweden
[3] Columbia National Fisheries Research Laboratory, U.S. Fish and Wildlife Service, Columbia, MO 65201

The overall objective of the use of pattern recognition in environmental problems is to identify, categorize or classify samples based on chemical data describing the samples. A number of pattern recognition methods are available for application to measured chemical data. Although these methods can be used to classify single compounds or the components of complex mixtures, they sometimes differ considerably in the way in which the classification rules are derived and applied. The SIMCA (SImple Modelling by Chemical Analogy) method is unique in having been developed specifically for application to chemical data and it has been shown in a number of studies to work very well. Its advantages are discussed.

The potential of modern chemical instrumentation to detect and measure the composition of complex mixtures has made it necessary to consider the use of methods of multivariable data analysis in the overall evaluation of environmental measurements. In a number of instances, the category (chemical class) of the compound that has given rise to a series of signals may be known but the specific entity responsible for a given signal may not be. This is true, for example, for the polychlorinated biphenyls (PCB's) in which the clean-up procedure and use of specific detectors eliminates most possibilities except PCB's. Such hierarchical procedures simplify the problem somewhat but it is still advantageous to apply data reduction methods during the course of the interpretation process.

A method that has been used with increasing success is the SIMCA method of pattern recognition (1). This method is extremely powerful when applied to data on complex mixtures, and a number of reports on such applications have recently appeared (2, 3, 4).

Steps in a Pattern Recognition Study

Pattern recognition methods are usually applied in discrete steps, which are outlined here. It is assumed that the chemical measurements

0097–6156/85/0292–0243$06.00/0

that characterize the samples in a study are relevant to the formu-
lated problem.

Steps

1.   Establish training sets.
2.   Derive classification rules.
3.   Select features.
4.   Refine classification rules.
5.   Classify unknowns.
6.   Review results graphically.

Various methods work in similar ways with regard to some of the
steps but the methods may differ in very significant and critical ways
in other steps. In some ways the SIMCA method is unique when viewed
in the context of these steps.

Establishment of Training Sets. The training sets are the samples or
compounds that are to serve as the basis of classification. It is
assumed that this set and its associated data are representative of
the class of samples that are to be categorized.
    The process of setting up training sets is somehwat arbitrary in
the sense that it is based on the experience and knowledge of the
person or persons conducting the analysis. This step is not a function of the method of analysis being used. It is important, however,
that an analytical chemist who is intimately familiar with the
instrument and with sample behavior be involved at this point.
    Once the training sets have been established, it is necessary to
obtain data on them relevant to classification of subsequent samples.
These data are the basis of the classification rules to be derived.
These samples of unknown class assignment are known as the test
samples or collectively as the test set. The training set(s) and test
set are tabulated with their data, as in Figure 1.
    In matrix notation, the data describing the samples can be ex-
pressed as a vector, as the Equation (1), and each sample, is then

$$X_k = \{x_1, x_2, x_3, \ldots x_i \ldots x_p\} \tag{1}$$

represented as a point in p-dimensional space. When the data for the
training sets are projected into variable space, the classes, ideally
cluster as in Figure 2.

Derivation of Classification Rules

Up to this point the methods of classification operate in the same
way. They differ considerably, however, in the way that rules for
classification are derived. In this regard the various methods are of
three types: 1) class discrimination or hyperplane methods, 2) distance methods, and 3) class modeling methods.
    In the class discrimination methods or hyperplane techniques, of
which linear discriminant analysis and the linear learning machine
are examples, the equation of a plane or hyperplane is calculated that
separates one class from another. These methods work well if prior
knowledge allows the analyst to assume that the test objects must

variable

sample	x_1	x_2	$\cdot$	$\cdot$	x_i	$\cdot$	x_p
1	x_{11}	x_{12}			x_{1i}		
2					$\cdot$		
3					$\cdot$		
$\cdot$					$\cdot$		
k	x_{k1}	$\cdot$	$\cdot$	$\cdot$	x_{ki}		
$\cdot$							
n							

Figure 1. Data matrix for a pattern recognition study.

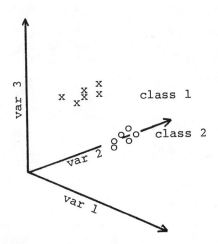

Figure 2. Graphical representation of 2-classes of pattern recognition data in 3-dimensions.

belong to one of the two classes. The possibility that the test set samples are members of neither of the training sets is not allowed by these methods. It is also necessary that the number of objects be much greater than the number of variables.

The distance methods operate differently. The classification of a test set member is based on the class assignment of the samples in the training set nearest to the unknowns. The type of distance used can differ but is usually the Euclidian distance, and the number of nearest neighbors is selected in advance. Usually the 3 to 5 nearest neighbors are selected and the possibility that the unknown may not be represented in the training sets is allowed.

Only one class modeling method is commonly applied to analytical data and this is the SIMCA method ($\underline{1}$) of pattern recognition. In this method the class structure (cluster) is approximated by a point, line, plane, or hyperplane. Distances around these geometric functions can be used to define volumes where the classes are located in variable space, and these volumes are the basis for the classification of unknowns. This method allows the development of information beyond class assignment ($\underline{5}$).

The class models are of the form of Equation 2, which is a

$$x_{ki} = \overline{x}_i + \sum_{a=0}^{A} t_{ka}\, k_{ai} + e_{ki} \tag{2}$$

principal components model. For A=0 the samples in a class are identical and the class is represented by a point in space; for A=1 it is represented by a line, and for $A \geq 2$ by a plane or hyperplane.

The objective of principal components modeling is to approximate the systematic class structure by a model of the form of Equation 2. This is shown diagramatically below in Equation 3. Here X is the

$$\boxed{X} \;-\; \left| \frac{\underline{}}{b} \right. \;=\; \boxed{E} \tag{3}$$

t

data matrix for a training set. The vector product of the t's (principal components) and b's (loadings), represent the systematic part of the matrix and E represents the residual matrix.

In this example, two principal components are arbitrarily selected. More or fewer may be necessary, and this is a function of a predetermined stopping rule for extraction of principal components from X. In SIMCA method, a cross validation technique ($\underline{7}$) is used.

SIMCA uses the NIPALS (Nonlinear Iterative PArtial Least Squares) algorithm for principal component abstraction ($\underline{6}$). Due to the simplicity of the algorithm and the ease of programming it for use on small computers, a short discussion of the NIPALS algorithm is presented here. The NIPALS procedure works as in Figure 3. One arbitrarily selects a normalized vector b. The product X b' is a n x 1 vector. The product of this vector in transpose with X gives a 1 x p vector. Normalized, this vector can be used to update b in the first multiplication. This is continued iteratively until t and b converge, usually within about 25 interactions. The product t   b is then substracted from X and the process continued with the matrix E until a stopping point is reached. This method has the advantage of not requiring matrix inversion for calculation of the principal components.

Feature Selection

Feature selection is the process by which the data or variables important for class assignment are determined. In this step of a pattern recognition study the various methods differ considerably. In the hyperplane methods, the strategy is to begin with a block of variables for the classes, calculate a classification function, and test it for classification of the training set. In this initial phase, generally many more variables are included than are necessary. Variables are then detected in a stepwise process and a new rule is derived and tested. This process is repeated until a set of variables is obtained that will give an acceptable level of classification.

This approach to feature selection leads to a set of descriptors that are optimal for class discrimination. These variables may or may not contain information that describes the classes.

In SIMCA, a class modeling method, a parameter called modeling power is used as the basis of feature selection. This variable is defined in Equation 4, where S_i is the standard deviation of a vari-

$$MPOW = 1 - S_i/S_{i,y} \tag{4}$$

able after it is fitted to Equation 2 and $S_{i,y}$ is the standard deviation before it is fitted to the model. This parameter is a measure of how well the variable contributes to the systematic class structure; its values are in the range $0 \le MPOW \le 1$. Variables with low MPOW, which are considered noise, are deleted; those with high MPOW are retained.

This criterion for selection of features leads to a set of descriptors that contain optimal information about class membership as opposed to information about class differences.

Model Refinement

After the feature selection process has been carried out once by the SIMCA method, it is necessary to refine the model because the model may shift slightly. This refining of the model leads to an optimal set of descriptors with optimal mathematical structure.

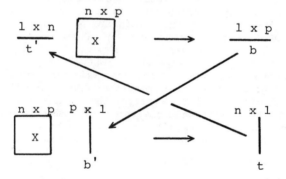

Figure 3. NIPALS algorithm for extraction of principal
components from a data matrix.

Classification of Unknowns

Class assignment, by the methods of classification discussed earlier, differs considerably. In the hyperplane methods, a plane or hyper plane is calculated that separates each class, and class assignment is based on the side of this discriminant plane on which the unknown falls. The limitation of this approach is that it requires prior knowledge (or an assumption) that the unknown be a member of one of the classes in the training sets.

In the distance methods, class assignment is based on the distance of the unknown to its k-nearest neighbors; since the distances of the training set objects from each other are known, one can determine whether an unknown is not a member of the training sets.

Since SIMCA is a class modeling method, class assignment is based on fit of the unknowns to the class models. This assignment allows the classification result that the unknown is none of the described classes, and has the advantage of providing the relative geometric portion of the newly classified object. This makes it possible to assess or quantitate the test sample in terms of external variables that are available for the training sets.

Graphical Presentation of Results

This aspect of data analysis is somewhat neglected, as it is associated more with the interpretation of results than with the analyses.

It is important to be able to view the structure of the data for the classes. This is done in a variety of ways depending on the analytical methods. The graphical technique most commonly used is that of plotting eigenvectors or principal components. SIMCA uses this method and software has been developed for three-dimensional color display of principal components data. Other plotting techniques are also used in SIMCA.

The SIMCA method of pattern recognition is in a comprehensive set of programs for classification, and we have discussed how it works in this regard. Classification problems represent only a few of types of problems that can be solved with this approach.

Literature Cited

1. Wold, S. *Pattern Recognition* 1976, , 127-139.
2. Dunn, W. J. III; Stalling, D. L.; Schwartz, T. R.; Hogan, J. W.; Petty, J. D.; Johansson, E.; Lindberg, W.; Sjostrom, M. *Analysis* 1984, 12, 477-485.
4. Lindberg, W.; Persson, J.-A.; Wold, S. *Anal. Chem.* 1983, 55, 643-648.
5. Albano, C.; Dunn, W. J. III; Edlund, U.; Johansson, E.; Norden, B.; Sjostrom, M.; Wold S. *Anal. Chim. Acta Comp. Tech. Optim.* 1978, 2, 429-443.
6. Wold, S.; Albano, C.; Dunn, W. J. III; Esbensen, K.; Hellberg, S.; Johansson, E.; Sjostrom, M. In "Food Research and Data Analysis"; Martens, H.; and Russwurm, H.; Eds., Science: London, 1983; pp. 182-189.
7. Wold, S. *Technometrics* 1978, 20, 397-406.

RECEIVED July 17, 1985

A Quality Control Protocol for the Analytical Laboratory

Robert R. Meglen

Center for Environmental Sciences, University of Colorado at Denver, Denver, CO 80202

A modified Youden two sample quality control scheme is
used to provide continuous analytical performance surveillance. The basic technique described by other workers
has been extended to fully exploit the graphical identification of control plot patterns. Seven fundamental
plot patterns have been identified. Simulated data were
generated to illustrate the basic patterns in the systematic error that have been observed in actual
laboratory situations. Once identified, patterns in the
quality control plots can be used to assist in the diagnosis of a problem. Patterns of behavior in the systematic error contribution are more frequent and easy to
diagnose. However, pattern complications in both error
domains are observed. This paper will describe how patterns in the quality control plots assist interpretation
of quality control data.

Analytical chemists performing routine analyses have long recognized
the need for a method of monitoring the performance of their analytical procedures. Quality control techniques have varied in sophistication from simple subjective evaluations by an experienced analyst who
knows that the results "don't look right" to more rigorous statistical protocols. Electronics and microprocessor advances have made
automated instrumentation widely available and have revolutionized
the modern analytical laboratory. Instrumental advances have made it
possible to generate massive quantities of data with minimal operator
attention in a fraction of the time once required for much smaller
efforts. In some cases the analyst's role has been reduced to feeding
samples to the instrument and retrieving the final report from an
output device. Many instrumental parameters are now under computer
control and the analyst's interaction with the measurement process is
minimized. In this analytical environment the need for a quality
control program is especially critical since anomalous instrument
performance may not be detected before several samples have been
"analyzed".
    The instrumental revolution has also lead to a data affluence
previously unrealized. Multielement techniques capable of simul-

0097-6156/85/0292-0250$06.25/0

taneous determination of dozens of chemical species has increased the
analyst's burden to perform determinations previously absent from
his/her repertoire. The analyst must now validate sample preparation
techniques for multiple species and maintain performance surveillance
for all species being reported. It may be necessary to maintain
quality control records for dozens of species. This can now be done
efficiently using automatic acquisition of quality control parameters
and computer generated summaries. By exploiting the advantages of
automated data acquisition the analyst is free to devote more time to
those aspects of chemical analysis that require human intuition,
experience, and interpretive skills.

The analyst's task in ensuring accurate and precise analyses
should extend beyond the laboratory to the sampling process. It is
not unusual for large numbers of samples to be collected by individuals who have little or no experience with the difficulties that
attend the selection of representative samples or with the steps
needed to preserve the sample after collection. Environmental
samples, for example, have great compositional variability and they
are often collected with little regard for the factors that determine
the validity of the final analytical result. Therefore it is essential that the analyst be involved in the design and implementaion of
the sampling program. An effective quality control scheme should
include attention to all aspects of the system being studied; sampling, sample treatment, sample preparation and analysis. The Association of Official Analytical Chemists (AOAC) has published a list
of concerns of the analyst in providing accurate and precise analyses
(1). They are listed here because they provide a useful background
for the quality control method that will be described here.

1. The method of choice must be demonstrated to apply to the
   matrices and concentrations of interest.

2. Critical variables should be determined and the need for controls
   emphasized.

3. Quality control samples must be identical and homogeneous so that
   the analytical sampling error is only a negligible fraction of
   the expected analytical error.

4. If the analyte is subject to change (bacterial, air oxidation,
   precipitation, adsorption on container, etc.) provisions should
   be made for its preservation.

5. Practice samples (for method validation) of known and declared
   composition should be available.

While the importance of adequate sampling design cannot be overem-
phasized we will not examine this aspect of quality control here.
Instead we will examine a few laboratory practices that are important
to the quality of the laboratory phase of the analytical result.

Quality Conrol - General Philosophy

The design of a total quality control protocol is based upon two
fundamental components; validation of the method and continuous

performance surveillance. The analyst first prepares a list of candidate methods. During the initial evaluation procedure samples of
the matrix of interest and "known" composition are treated according
to published or established procedures. Methods that provide promising results are then refined through a method development process
that adapts the method to particular matrix characteristics. An
optimized method is then used to analyze samples of "known" composition and of a matrix similar to anticipated unknowns. A standard
reference material (of similar matrix characteristics, if available)
is also analyzed to assess the accuracy of the proposed method. Once
an accurate method has been developed routine analyses are begun. A
sample of known composition (a quality control sample) and standard
reference material are periodically analyzed with the routine unknowns to ensure that the method's routine application continues to
provide adequate results. This procedure of re-analyzing a single
sample provides a time link for accuracy between the method validation period and routine work. Precision is assessed by performing
replicate analyses on both Q.C. samples and unknowns. When erroneous
results are detected, routine analyses are suspended and additional
method development work may be required to improve the procedure.
Figure 1 illustrates the general scheme that is common to most
quality control procedures. The specific protocol presented in this
paper is based on this scheme but it has been modified to enhance the
ease with which one may assess analytical performance and diagnose
problems. Since the first step in any Q.C. process, validation, is
the key to overall performance we will briefly examine the procedures
used to assess accuracy and precision.

Phase One: Validation, Accuracy, and Precision

Accuracy is a measure of how close a measurement is to the "true"
value. While it is impossible to determine absolute accuracy it is
possible to obtain an accuracy estimate using several techniques.

Certified Reference Materials.   Certified Reference Materials are
materials whose properties have been guaranteed or certified by
recognized bodies. The certified analyses of these materials can be
used as an estimate of the "true" value for assessment of accuracy.
The United States National Bureau of Standards (NBS) provides an
inventory of various materials whose compositions (and properties)
have been measured using definitive and reference methods. These
materials, Standard Reference Materials (SRM's), when used in conjunction with reference methods, i.e., one of demonstrated accuracy,
make it possible to transfer accuracy between measurement protocols.
    Other classes of reference materials now in existence include
secondary reference materials. These are materials produced commer-
cially for reference purposes, but whose guarantee rests soley with
the producer. "Analyzed" materials such as geological materials
obtained from the United Staes Geological Survey, represent test
samples that complement the variety available from the previously
mentioned sources. However, the "accepted" analyses reported for
these materials are based upon consensus values obtained from large
scale interlaboratory collaborative tests (round robins). Analysis of
these materials can provide a useful means of comparing performance
with other laboratories, but it does not ensure accuracy. In addi-

Conventional scheme:

provides accuracy time-linked by
repeated analysis std. ref. materials

precision - replicate analysis

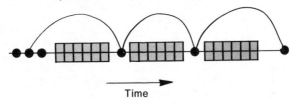

Time

Figure 1. Diagram showing conventional scheme for linking accuracy over time by periodic analysis of reference material.

tion, one must be cautious of assessing accuracy by comparing results
with consensus values because they are often computed incorrectly in
the literature. The quoted values may be based upon averages of all
methods without regard to systematic bias that characterizes some
methods of analysis.

The selection of an appropriate reference material should be
based upon the availability of a matrix that is similar to the anticipated routine unknowns. Similarity of chemical matrix and analyte
concentrations is particularly important when attempting to assess
accuracy of a method that requires destructive sample preparation.

Independent Methods. In the absence of appropriate certified
reference materials one may have to rely upon in-house materials that
can be analyzed by independent methods (other than the candidate
method). These independent methods should include a reference method
and other methods that utilize different physical/chemical principles
for analyte quantification. Reference methods are generally arrived
at by concensus following extensive accuracy testing by a large
number of laboratories. The American Society of Testing Materials
(ASTM) is one of the largest compilers of reference methods. Additional information on the use of reference methods may be found in a
paper by Cali and Reed (2).

Collaborative Testing. A second approach to assessing accuracy,
when no certified reference material is available, may be used in
conjunction with analysis by independent methods and in-house
materials. Sample exhanges with other laboratories can help establish
the existence or absence of systematic errors in a method. Collaborative tests are most useful in this regard when some of the participating laboratories use different sample preparation and quantification. The utility of independent analysis methods and comparisons between destructive and non-destructive analysis is again
emphasized here.

Referee Laboratories and Spike Recovery Testing. Outside
laboratories, with demonstrated performance records, can be used to
evaluate the suitability of a candidate method when none of the other
accuracy testing options is feasible. However, This technique
provides a very weak form of accuracy assessment. Indeed, it provides
a comparability check, not an accuracy measure. Similarly, spike
recovery tests provide only weak evidence of method accuracy. Quantitative spike recovery only indicates that the added form of the
analyte was recovered. If the added form responds differently toward
sample preparation or detection the utility of spike recovery testing
remains doubtful.

Accuracy is an expensive commodity. It involves exhaustive
testing of the candidate method. Thorough delineation and careful
control of analytical variables is essential to accurate analyses.
The expenditure of substantial effort in the early stages of method
development will be more efficient and less embarassing than later
corrective work.

Precision is a measure of the reproducibility of a given result.
The role of precision in demonstrating a method's accuracy has not
been addressed. However, a clear understanding of the Q.C. method
being presented here requires that we briefly examine a few basic
features of measurement errors.

There are two types of errors associated with any chemical analysis; **systematic** (determinate) and **random** (indeterminate). Inaccurate results, consistently higher or lower than the "true" value, occur when systematic errors are present. Systematic errors tend to have the same algebraic sign and usually arise from erroneous calibration, intrumental drift, loss of analyte or contamination during sample preparation, failure to account for blank or background effects, etc. Through adequate testing procedures it is possible to determine the magnitude and source of this type of error (hence the term "determinate error"). The validation phase of methods development is designed to eliminate this source of error. Occasionally systematic errors appear after the method has been in use for some time. **An effective quality control scheme should permit early detection of systematic error and assist the analyst in diagnosing its cause.**

Random or indeterminate errors arise from a large number of minute variations in materials, equipment, conditions, etc. If these factors are truly independent the sum of all fluctuations will result in small positive or negative errors that have the highest probability of occurrence. Very large positive or negative deviations are less probable. These errors are inherent in any measurement technique that is based on a continuous interval scale such as; reading peak heights or analog meters, determining mass, etc. While careful control of experimental variables can minimize the magnitude of these errors, they are always present. These errors define the precision of the measurements and establish the detection limit of the procedure. While the distribution of these random errors need not be normal, normal distributions are observed for most analytical chemistry measurements. The derivations that follow are based on the assumption of normality. **An effective quality control scheme should permit early detection of any change in the magnitude of random errors and assist in diagnosing its cause.**

Phase Two: Surveillance Monitoring

The second phase of a <u>total</u> quality control scheme continues beyond the initial validation phase described earlier. It consists of systematic performance monitoring and provides a time-link to the accuracy established during the validation phase. Quality control monitoring requires continuous surveillance to determine the onset of systematic errors and the appearance of large random errors that affect precision. The ability to distinguish between random and systematic error contributions to measurements is an important prerequisite to problem diagnosis. The technique described by W. J. Youden (3,4) was designed to identify and separate systematic and random errors that occur among laboratories participating in **inter** laboratory tests. The method has been modified by King (5) and extended by Meglen (6) for use within a single laboratory, **intra** laboratory testing. In this modification the results obtained from day-to-day are treated as if they were obtained in different laboratories. Two different plotting techniques are used to monitor the analytical performance. Following a brief description of the mathematical basis for this approach we will examine several example plots. A detailed derivation of the within run and between run variances is given in reference 5.

Assume that a single sample is split into two portions labeled A and B. A quantitative determination of some sample constituent should yield the "true" value X plus any systematic and random error contributions.

$$A = X + S + R \tag{1}$$

$$B = X + S + R' \tag{2}$$

Where S is the systematic error (bearing the same algebraic sign and having the same magnitude for each sample), and R and R' are the random errors (bearing potentially different algebraic signs and having different magnitudes for each sample). We have assumed that the random error contributions for each sample have equal probability of being either positive or negative; i.e., they are normally distributed and independently expressed. The sum of the results obtained for both splits will yield a number T that has twice the true value and twice the systematic error of one sample.

$$T = (A + B) = 2X + 2S + R + R' \tag{3}$$

Since the random error contributions, R and R', have identical distributions symmetric about zero, and with expectation of zero; an average value of T, based on a large number of observations will have a very small component from averaging of R and R'.

When the difference D between the results A and B is computed the systematic errors, which have the same magnitude and sign, will cancel. This leaves the difference of the two random error components, which do not necessarily cancel for a particular pair.

$$D = (A - B) = R - R' \tag{4}$$

By plotting the sum T and difference D in time ordered sequence the variation of random and systematic errors can be monitored between analytical runs.

The procedure used for day-to-day monitoring utilizes a single real sample (usually a composite of previously analyzed samples) split into two aliquots labeled A and B. These samples are carried through the analytical procedure together with the unknowns.

Graphical Display

The primary purpose of any quality control scheme is to identify ("flag") significant performance changes. The two-sample quality control scheme described above effectively identifies performance changes and permits separation of random and systematic error contributions. It also permits rapid evaluation of a specific analytical result relative to previous data. Graphical representation of these data provide effective anomaly detection. The quality control scheme presented here uses two slightly different plot formats to depict performance behavior.

Youden described a plotting protocol that depicts the relative positions of individual runs on two samples. Consider the hypothetical case where an analytical method has been perfected and no sys-

tematic error is present. The determinations on two samples, A and B,
would then have the following deviations due to inherent random
error: i.e., both slightly high, both slightly low, and one slightly
high and one slightly low.

Sample A	Sample B
+	+
−	−
+	−
−	+

All four possibilities would be equally likely in an accurate method.
The results from a series of paired determinations on samples A and B
may be plotted on two axes. For any given analytical run the result
of the A determination may be plotted against the result of the B
determination. For a large number of runs, a vertical line may be
drawn through the average of the B results; a horizontal line may be
drawn through the average of the A results. The plot is thereby
divided into four quadrants. The quadrants correspond to the four
outcomes enumerated above; upper right, lower left, upper left, and
lower right respectively. If the only source of error is truly random
all four quadrants should be equally populated. (See Figure 2a.)

To gain further insight regarding the distribution of points on
this type of plot we shall consider the hypothetical case where no
random error exists. All errors are systematic and each determination
has associated with it either a high bias or a low bias. When these
results are plotted on the quadrant axes the points would lie in the
upper right (++) or lower left (−−) quadrants. If the systematic
error for both samples were equal the plotted points would describe a
straight line with unit slope (45 degrees). See Figure 2b.

Actual experience shows that random errors can only be minimized, not eliminated; and a quadrant plot would generally appear as
shown in Figure 2c. Figure 2d shows an ellipse which is drawn to
enclose 95% of the results obtained in a hypothetical experiment that
exhibits minimal random errors and small systematic errors. The
ellipse's major axis is equal to two standard deviations (obtained
from the Total variance; i.e., Between-run). The minor axis is equal
to two times the random error standard deviation (obtained from the
random error component; i.e., Within-run variance). The circle is
drawn to enclose 95% of the random error results. Thus, points found
in the region between the circle and the ellipse have a high probability of being the result of systematic error. (Similar elliptical
patterns would be observed in the absence of systematic errors if R
and R' were bivariate normal with different variances. However, since
both Q.C. samples, A and B, are the same material, we assume that
their variances will be equal.) The application of the Youden method
to intralaboratory evaluation derives its utility by incorporating
time as a variable. By connecting the points on a quadrant plot in
time-ordered sequence it is possible to identify time dependent
variations within the random and systematic error domains. Detailed
examples and interpretive aids that exploit this feature will be
provided later. We will refer to this type of plot as Q-plots for the
remainder of the discussion.

A second type of time dependent plot provides complementary
information for performance evaluation. Results of the determinations
on A plus B (Totals, T) and A minus B (Differences, D) are each
plotted in time ordered sequence (or linear time scale) to facilitate
detection of time dependent patterns or trends. When the analytical
procedure is under control systematic errors are eliminated and
random errors are minimized. The resultant T versus time plots show a
linear distribution of points with zero slope. The scatter is determined by the magnitude of the nominal precision and does not change
with time. Similarly the points on plot of D versus time should be
normally distributed about zero and the dispersion (standard deviation) is constant with respect to time. Changes in T-plots, in the
absence of concomitant changes in D-plots, indicate changes in
analytical procedures that contribute to systematic errors. Changes
in D-plots imply that controls on the random error sources have
failed. It is possible to set statistically based control limits that
signal the "out of control" condition and mandate suspension of
routine analytical work until remedial actions are taken. By using Q,
T, and D-plots it is also possible to gain insight for problem diagnosis. Diagnostic techniques will be discussed later.

The Quality Control Protocol Design

The main purposes of a quality control scheme are to provide accuracy
and precision monitoring. Since accuracy is established during the
method development phase, a time link to that process is essential.
Long term monitoring is provided by replicate analyses over the
duration of the program. Therefore, a sufficient quantity of the
quality control sample should be available. It should be homogeneous,
stable, and duplicate the "unknown's" sample matrix. The analyte
concentrations should also represent the real sample range. This
ensures that systematic error resulting from sample matrix effects
will be detected if control measures fail. While SRM's afford the
advantage of "known" composition, they may not be available in sufficient quantity for long term programs and they are not available in a
wide variety of matrix types. A homogenized composite of the routine
samples has the desired representative sample characteristics.
    The Q.C. sample should be stable with respect to physical,
chemical and biological change. Trace element constituents of aqueous
samples are susceptible to biological conversion, air oxidation, and
absorption on container surfaces. Aqueous quality control samples
that have high concentrations, near saturation, should be protected
from temperature fluctuations that may cause precipitation and redis-
solution. Caution should be exercised when compositing aqueous
samples for use as a Q.C. sample since disparate samples may lead to
chemical reactions that achieve equilibrium slowly. Precipitation of
analyte may occur over an extended time period. The Q.C. protocol
described here will detect these Q.C. sample changes if they occur,
but it will not distinguish them from analytical performance or
instrument changes.
    The ultimate Q.C. protocol should anticipate the potential error
sources in the whole analytical procedure. Destructive analyses that
require extensive treatment for sample decomposition can lead to
large analytical errors. Solid sample fusions and digestions should

be monitored. This means that a Q.C. sample should be carried through the entire analytical procedure. If instrument performance is the principal concern, a composite of several prepared samples will provide a convenient means of monitoring the detection and quantification step in the analytical process. Complete segregation of error sources requires a multi-level approach to Q.C. protocol design. The simplified two sample scheme described here may not be sufficient for all monitoring purposes. The analyst should examine the need for applying an analysis of variance (ANOVA) protocol if multi-level information is desired.

The two sample quality control protocol described here is shown schematically in Figure 3. The repeated analysis of the A and B pair provides the accuracy time link between method validation and routine work. Since A and B are aliquots of the same sample they provide data for computing precision (within and between analytical runs). Additional real sample replicates may be added at the discretion of the analyst. However, the present scheme minimizes the time consuming requirement for performing numerous sample replicates. If the Q.C. sample is truly matrix and concentration representative the precision computed from its replication will provide a valid estimate of the unknown sample's analytical reproducibility. When significant precision excursions have been identified from Q.C. plots more extensive testing may be indicated. The purpose of the Q.C. procedure is to provide continuous performance monitoring. The utility of the Q.C. plots for identifcation and diagnosis of analytical problems may be exploited only through frequent examination. The results of these evaluations facilitate the dynamic interaction between analytical methods development and routine work.

Selecting the placement of Q.C. samples within the anaytical run depends upon the purpose of the Q.C. program. While random placement is statistically justified, it may not provide sufficient diagnostic information. If instrumental drift is an important concern (as it is in many automated, operator unattended techniques) the two Q.C. samples should be spaced at intervals that are appropriate to detect the anticipated drift. Placement near the beginning and end of the analytical run has been been beneficial in detecting instrumental drift. By bracketing groups of routine samples with Q.C. samples it is easy to identify specific samples that require re-analysis.

The number of Q.C. pairs relative to the number of routine samples also depends upon the judgement of the analyst. Short term instrumental fluctuations require frequent use of Q.C. pairs. Practical considerations such as autosampler capacity and the number of samples that an operator can handle affect the Q.C. sampling frequency. An appropriate guideline of one Q.C. pair per twenty routine samples has been used effectively in most operations. A separate Q.C. sample pair should be included in every "run". (A "run" may be defined as any separate application of the analytical procedure characterized by a change in calibration status, operator, reagent lot, or instrument operational characteristics; e.g. on-off cycle, tuning, cleaning, maintenance, etc.)

Two aspects of Q.C. sample labelling require discussion. 1) Which sample should be "A" and which "B"? 2) Should samples be analyzed "blind"? Labelling samples "A" and "B" is merely an operational convenience. The labels are only used to prepare Q.C. plots and compute sums, differences, and statistics. If the first Q.C.

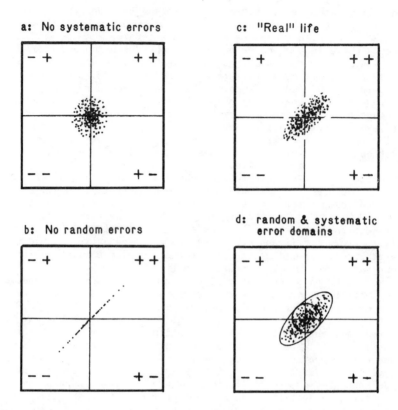

Figure 2. Plots showing location of measured values with various systematic and random error contributions.

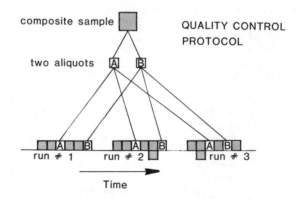

Figure 3. Schematic diagram showing the use of two Q.C. samples for long-term monitoring of systematic errors.

sample analyzed is always "A" and the second is always "B". instrumental drift within the analytical run is easily detected. This
is true because any D-plot should have points normally distributed
about the difference of zero. Any non-random D-plot distribution may
be interpreted as within-run (short term) systematic error, i.e.
drift. The Q-plot will also show the systematic difference between
the two determinations since the running mean of A's and B's determines the location of the plot's horizontal and vertical axes. Thus,
assymetric axis location will reflect systematic bias introduced
during the run.

The selection of labelling need not affect the "blind" nature of
the analysis since Q.C. samples do not have to be identified until
analyses are completed. Treating the Q.C. samples in "blind" fashion
is often important to ensure that they do not receive special treatment. These samples are used as surrogate replicates for real samples
and are used to evaluate method performance in lieu of routine unknown sample replicates. Therefore, they must not receive special
operator attention or handling. However, the "blind" requirement may
be relaxed when sample preparation has been minimal or well controlled, or when automated instrument performance is the sole subject of
scrutiny. It may be argued that "blind" labelling is unecessary even
when the detection device is under human operator control since any
attempt to "adjust" the determination of either Q.C. sample to match
its pair mate will be expressed as an anomalous difference D.

Patterns in the Systematic Error

Simulated data were generated to illustrate the basic patterns in the
systematic error that have been observed in actual laboratory situations. The magnitude of the effects have been exaggerated so that the
essential features of the interpretation may be illustrated. These
hypothetical data were computer generated such that the "true" value
of A and B should be 100 units. The random error contribution was
generated such that each simulated measurement was taken from a
normally distributed error population with a standard deviation of 5
units. Figures 4 through 10 illustrate the simplest patterns that are
commonly observed in the laboratory. Additional combinations (28) of
patterns in both systematic and random error components described by
Meglen (6) are not shown here.

Commentary on Example Plots. The following commentaries describe
characteristics of typical patterns observed in the laboratory.

NONE.
        (Figure 4) No systematic error is present. The only error con-
    tribution is the result of random deviations in the results
    obtained for the two quality control samples, A and B.
                            Q Plot:
    The shape of the distribution is circular with lines of equal
    lengths at random angles. There is an equal distribution of
    points among the four quadrants. (Normally distributed; dense in
    the center, sparse in the outer region.) No systematic error is
    detected.
                            T Plot:
    Spurious high and low points corresponding to small errors in

the D plot suggest possible systematic errors. However, this is
also consistent with normally distributed errors.
D Plot:
Normally distributed random errors are shown with no apparent trend.

Total   RSD = 5.1 %
Random   RSD = 4.8 %
Systematic RSD = 1.5 %

FREAKS
        (Figure 5) Systematic error is present, but it does not follow a
simple functional relationship with time. This case is simulated
by the occurence of a large systematic error component (greater
than three standard deviations from the mean) which appears
without warning.
        Freaks in the T charts generally occur simultaneously with
freaks in the D plots. They are generally caused by sudden
introduction of bias such as; sample contamination, loss of
analyte, calibration standards or reagents gone bad, careless-
ness or failure to control operating parameters. Some freaks are
to be expected in any stochastic process. However, frequently
reoccuring anomalies suggest a systematic source for the bias.
Careful scrutiny of operations often reveals an underlying
pattern that leads to recurrent freaks (e.g all freaks are
produced by a single operator; or all freaks occur on a particular work day, implying an environmental factor.)
Q Plot:
An elliptical distribution indicates that systematic error is
detected. Long line segments at 45 degrees deviate from an
otherwise random direction and equal placement among the four
quadrants. Anomalous points are well outside of the ellipse in
the systematic error quadrants. The points are otherwise normally distributed.
T Plot:
No systematic pattern appears. Only spurious high and low points
are seen.
D Plot:
An absence of large random error contribution corresponding to
anomalous points in the T plot shows that they are in the systematic error domain. (This is more readily seen in the Q plot.)

Total   RSD = 6.7 %
Random   RSD = 3.9 %
Systematic RSD = 5.5 %

SHIFT
        (Figure 6) The systematic error contribution obeys a step func-
tion. The absence of any systematic bias during the early time
period is followed by a sudden appearance of a large constant
systematic error (either positive or negative).
        Sudden shifts to lower or higher values in the T plot are
generally operator related. Different operators may use slightly
different procedures that lead to bias. New reagent lots may
introduce systematic error through blank contamination or different potency. Sudden undocumented environmental events may

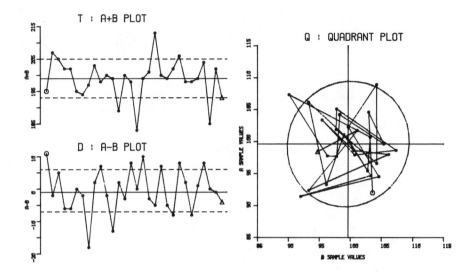

SYSTEMATIC ERROR PATTERN : NONE
RANDOM ERROR PATTERN : NONE

Figure 4. Example Q.C. plots showing no systematic error pattern. See commentary in text.

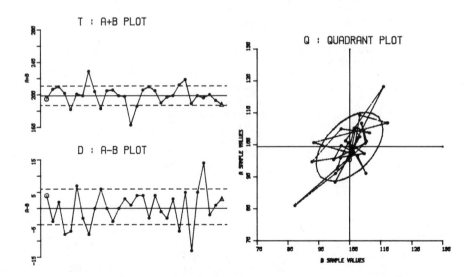

SYSTEMATIC ERROR PATTERN : FREAKS
RANDOM ERROR PATTERN : NONE

Figure 5. Example Q.C. plots showing the "FREAKS" systematic error pattern. See commentary in text.

change the operational characteristics of the instrument (physical abuse or movement of optically sensitive instruments by
janitorial staff has been encountered in some laboratories.)
                              Q Plot:
The elliptically shaped distribution shows that systematic
errors are detected. Two distributions are seen; each characterized by randomly directed short line segments. One long line
segment at 45 degrees between systematic error quadrants signals
the sudden systematic shift.
                              T Plot:
Points form a step function with a sudden increase (or
decrease).
                              D Plot:
No apparent pattern appears, and the points are normally distributed about zero.

Total   RSD = 8.4 %
Random   RSD = 4.6 %
Systematic RSD = 7.1 %

TREND
        (Figure 7) The systematic error contribution increases or
decreases monotonically with increasing time.
        Monotonic increases or decreases in the T plot are
generally related to changes in calibration standards, or the
Q.C. samples themselves. Failure to adequately preserve stored
standards or samples will lead to this pattern. Slow, constant
reagent degradation can also produce the TREND pattern.
                              Q Plot:
An elliptical distribution indicates that systematic errors are
present. Short line segments connect the points. They move
monotonically from one systematic error quadrant to the other.
There is an insufficient density of points in the middle of the
ellipse and in the random error quadrants for this to be a
normal distribution of errors.
                              T Plot:
The time sequence of points have non-zero slope; i.e., the
absolute value of T changes with increasing time.
                              D Plot:
Points are normally distributed about zero, no pattern appears.

Total   RSD = 12.2 %
Random   RSD = 3.5 %
Systematic RSD = 11.7 %

PLATEAU
        (Figure 8) The systematic error contribution increases or
decreases rapidly with time, but finally levels off to a constant value. This behavior is similar to, but occurs less
precipitously than, the step function exhibited by the SHIFT.
        Slow change to higher or lower values in the T plots, with
subsequent leveling off to a constant value characterize this
pattern. This behavior usually suggests the slow attainment of
an equilibrium value. Inadequately stabilized and equilibrated
calibration standards or Q.C. samples will lead to this pattern

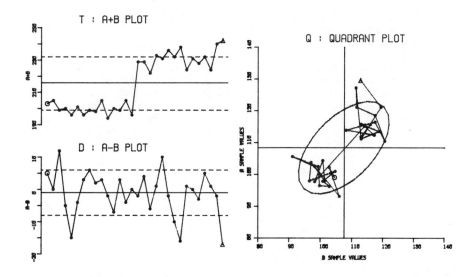

Figure 6. Example Q.C. plots showing the "SHIFT" systematic error pattern. See commentary in text.

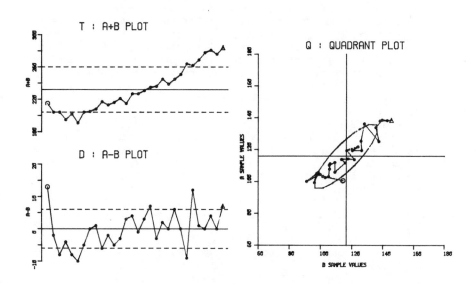

Figure 7. Example Q.C. plots showing the "TREND" systematic error pattern. See commentary in text.

in the T plot. Such patterns are seldom the result of instrumental changes unless the D plot shows a corresponding change in
magnitude.

Q Plot:

An elliptical distribution indicates that systematic errors are
present. The pattern is similar to Figure 7, labeled TREND;
however, the line segments are longer and too many points lie at
the ellipse extrema. This indicates two "level" regions in the T
plots (early and late). The time sequence connected points show
movement from the systematic low quadrant to the systematic high
quadrant.

T Plot:

The plot shows a rapid increase followed by a slower upward
trend which levels off later in the chart.

D Plot:

Normally distributed random errors are shown with no apparent
trend.

```
Total      RSD = 10.2 %
Random     RSD =  6.6 %
Systematic RSD =  7.8 %
```

CYCLE

(Figure 9) The magnitude of the systematic error contribution
changes continuously with time, but it follows a definite cyclic
pattern that repeats itself periodically. (This case is simulated here as a sine wave.)

Slow periodic variation of the T plot are usually the
result of uncontrolled environmental factors. Seasonal variations related to poor laboratory temperature control have been
frequently identified by this pattern. Instruments are seldom
sensitive to small ambient temperature changes. However, if an
instrument is operating near its suggested nominal operating
temperature; short term excursions from this temperature can
affect both accuracy and precision. CYCLES in the T plots are
usually accompanied by similar behavior in the D plots.

Q Plot:

Systematic error is evident in the clear ellipticity of the
distribution. The time ordered sequence shows a non-random
"walk" between systematic error quadrants. An excursion from one
systematic quadrant to another and a subsequent return is evident. The distribution is non-normal, with too few points in the
central region.

T Plot:

Sinusoidal fluctuation shows clear periodic behavior in the
systematic error domain.

D Plot:

Normal distribution of random errors is shown.

```
Total      RSD = 11.8 %
Random     RSD =  4.5 %
Systematic RSD =  9.8 %
```

BUNCHING

(Figure 10) The systematic error contribution undergoes several

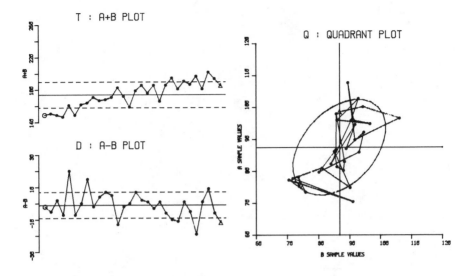

Figure 8. Example Q.C. plots showing the "PLATEAU" systematic error pattern. See commentary in text.

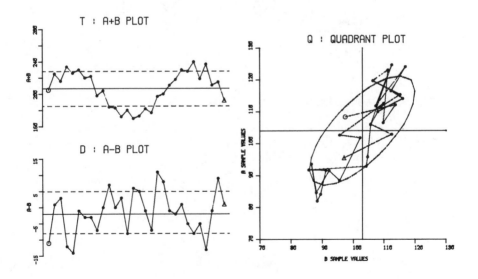

Figure 9. Example Q.C. plots showing the "CYCLE" systematic error pattern. See commentary in text.

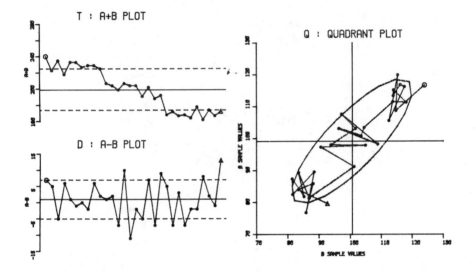

SYSTEMATIC ERROR PATTERN : BUNCHING
RANDOM ERROR PATTERN   : NONE

Figure 10. Example Q.C. plots showing the "BUNCHING" systematic error pattern. See commentary in text.

successive "quantized" magnitude changes. It is similar to
several successive SHIFTs.

The bunching pattern in T plots differ from cycles in two
respects; in bunching, the changes are precipitous, and they do
not have a characteristic repetition frequency. The sudden
systematic error shifts are due to apparently random events.
These events are most commonly associated with calibration
errors and/or operator technique. Rotation of laboratory personnel can produce this pattern if the individuals follow different
procedures. Operator related systematic errors can be detected
by plotting points with separate symbols for different
operators. Bunching may also appear when reagent lots are
changed.

Q Plot:
The elliptical distribution shows the presence of systematic
error. Three dense sub-clusters distinguish this behavior from
TREND. Note the long 45 degree segments along the systematic
error directions. Bunching is characterized by multimodal behavior with fairly dense sub-clusters. The SHIFT shown earlier
(Figure 6) is a special case of bunching, but a SHIFT generally
exhibits only one step change. True bunching behavior tends to
reoccur but without the predictable periodicity which characterizes CYCLES. CYCLES do not exhibit the long line segments
seen in this plot.

T Plot:
Randomly reoccurring stratification of results about different
localized means characterize this plot.

D Plot:
No apparent patterns appear and a random normal distribution is
seen.

Total   RSD = 13.2 %
Random   RSD = 4.2 %
Systematic RSD = 12.5 %

Conclusion

While methods validation and accuracy testing considerations
presented here have been frequently discussed in the literature, they
have been included here to emphasize their importance in the design
of a total quality control protocol. The Youden two sample quality
control scheme has been adapted for continuous analytical performance
surveillance. Methods for graphical display of systematic and random
error patterns have been presented with simulated performance data.
Daily examination of the T, D, and Q quality control plots may be
used to assess analytical performance. Once identified, patterns in
the quality control plots can be used to assist in the diagnosis of a
problem. Patterns of behavior in the systematic error contribution
are more frequent and easy to diagnose. However, pattern complications in both error domains are observed and simultaneous events in
both T and D plots can help to isolate the problems. Point-by-point
comparisons of T and D plots should be made daily (immediately after
the data are generated). Early detection of abnormal behavior reduces
the possibility that large numbers of samples will require re-
analysis.

Literature Cited

1. Schall, Elwin D.; "Collaboartive Study Procedures of the Association of Official Analytical Chemists."; Anal. Chem. vol. 50: #3, pp. 337A-340A, (1978).
2. Cali, J. Paul and Reed, William P.; "The Role of the National Bureau of Standards Standard Reference Materials in Accurate Trace Analysis."; pp41-63; in, Accuracy in Trace Analysis: Sampling, Sample Handling, Analysis., Proceedings of the 7th Materials Research Symposium; NBS Special Publ. #422, vol. 1&2; Ed. by Philip D. LaFleur., U.S. Department of Commerce, National Bureau of Standards; August, 1976.
3. Youden, W.J.; "Statistical Techniques for Collaborative Tests.", The Association of Official Analytical Chemists; Washington D.C., 1969 (Rev. 1973).
4. Youden, W. J.; "Interlaboratory Tests-Chapter Three"; NBS Special Publication #300, Vol. #1, 1969.
5. King, Donald E.; "Detection of Systematic Error in Routine Analysis."; pp.141-150; in, Accuracy in Trace Analysis: Sampling, Sample Handling, Analysis., Proceedings of the 7th Materials Research Symposium; NBS Special Publ. #422, vol. 1&2; Ed. by Philip D. LaFleur., U.S. Department of Commerce, National Bureau of Standards; August, 1976.
6. Meglen, Robert R.; "A Quality Control Protocol for the Analytical Laboratory"; Center for Environmental Sciences, University of Colorado at Denver, Special Publication (79 pages), 1983.

RECEIVED June 28, 1985

Statistical Receptor Models Solved by Partial Least Squares

Ildiko E. Frank[1] and Bruce R. Kowalski

Laboratory for Chemometrics, Department of Chemistry, University of Washington, Seattle, WA 98195

PLS (partial least squares) multiple regression technique is used to estimate contributions of various
polluting sources in ambient aerosol composition. The
characteristics and performance of the PLS method are
compared to those of chemical mass balance regression
model (CMB) and target transformation factor analysis
model (TTFA). Results on the Quail Roost Data, a
synthetic data set generated as a basis to compare
various receptor models, is reported. PLS proves to
be especially useful when the elemental compositions
of both the polluting sources and the aerosol samples
are measured with noise and there is a high correlation
in both blocks.

In the past few years, PLS, a multiblock, multivariate regression
model solved by partial least squares found its application in
various fields of chemistry (1-7). This method can be viewed as an
extension and generalization of other commonly used multivariate
statistical techniques, like regression solved by least squares and
principal component analysis. PLS has several advantages over the
ordinary least squares solution; therefore, it becomes more and more
popular in solving regression models in chemical problems.
    One of the current problems in environmental chemistry is how to
model the ambient aerosol composition, to reveal polluting sources,
to determine their contribution to the overall aerosol composition.
    In the past few years several receptor models were developed.
The basic assumption of these receptor models is that the ambient
airborne particle concentrations measured at a receptor can be appor-
tioned between several sources. In other words, each chemical
element concentration at the receptor is considered as a linear
combination of the mass fraction of the source contributions.
    The two most widespread statistical receptor models in the
literature are: regression model of chemical mass balance (CMB) (8)
and target transformation factor analysis (TTFA) (9). The questions
to be answered by the receptor models are:

[1] Current address: Jerll, Inc., Stanford, CA 94305

0097–6156/85/0292–0271$06.00/0

- How many sources are active?
- What is the chemical profile of these sources?
- What is the contribution of these profiles?

In this paper our goal is to introduce the PLS method, to discuss its properties, to compare it with the CMB and TTFA models and to demonstrate its performance on a well known synthetic data set.

The PLS Method

The PLS technique gives a stepwise solution for the regression model, which converges to the least squares solution. The final model is the sum of a series of submodels. It can handle multiple response variables, highly correlated predictor variables grouped into several blocks and underdetermined systems, where the number of samples is less than the number of predictor variables. Our model (not including the error terms) is:

$$y_{ik} = \sum_{j=1}^{NX} x_{ij} s_{jk} \qquad \begin{matrix} i = 1 \ \ldots \ NSAMP \\ k = 1 \ \ldots \ NY \end{matrix} \qquad (1)$$

where X is the predictor variable matrix of NSAMP samples and NX variables, S is the regression coefficient matrix of NX rows and NY columns and Y is the response variable matrix of NSAMP samples and NY variables. An extension of this model of several predictor blocks also can be solved by PLS (6), (7), but because only the two block model will be applied to the receptor model problem, this extension is not discussed here.

The variance in the predictor block is described by a set of latent variables U, which are linear combinations of the predictor variables. Similarly V is the latent variable matrix for the response block. These equations are called outer relationship

$$u_{i\ell} = \sum_{j=1}^{NX} x_{ij} \cdot a_{j\ell} \qquad \begin{matrix} i = 1 \ \ldots \ NSAMP \\ \ell = 1 \ \ldots \ NCOMP \end{matrix} \qquad (2)$$

$$v_{i\ell} = \sum_{k=1}^{NY} y_{ik} \cdot b_{k\ell}$$

where NCOMP is the number of the latent variables, which can be maximum NX. A and B are called the weight matrices. The latent variables are orthogonal to each other, similar to the principal components. To ensure the orthogonality of the latent variables, they are calculated from the residual matrices X' and Y'

$$x'_{ij} = x_{ij} - d_\ell \cdot u_{i\ell} \cdot c_{\ell j} \qquad \begin{matrix} i = 1 \ \ldots \ NSAMP \\ j = 1 \ \ldots \ NX \\ k = 1 \ \ldots \ NY \\ \ell = 1 \ \ldots \ NCOMP \end{matrix} \qquad (3)$$

$$y'_{ik} = y_{ik} - d_\ell \cdot u_{i\ell} \cdot b_{\ell k}$$

where C and B (same as the above weight matrix) are the orthogonal projections of the X and Y matrices on the submodel dU, respectively.

The weights A and B are calculated as correlations between the variables of one block and the latent variable of the other block in an iterative procedure. The iteration starts with an arbitrary latent variable vector of the response block (V). The

result after convergence is independent from the starting point,
unless one starts orthogonal to the solution, which is very unlikely.

$$a_{j\ell} = \sum_{i=1}^{NSAMP} x_{ij} \cdot v_{i\ell} \qquad j = 1. \ . \ .NX$$

$$b_{k\ell} = \sum_{i=1}^{NSAMP} y_{ik} \cdot u_{i\ell} \qquad k = 1. \ . \ .NY \qquad (4)$$

This calculation ensures maximum correlation between the latent
variables of different blocks. The submodel is:

$$v_{i\ell} = d_\ell \cdot u_{i\ell} \qquad \begin{aligned} i &= 1. \ . \ .NSAMP \\ \ell &= 1. \ . \ .NCOMP \end{aligned} \qquad (5)$$

where $\underline{d}$ is called the inner relationship coefficient.
The final model is the sum of the submodels:

$$y_{ik} = \sum_{\ell=1}^{NCOMP} d_\ell u_{i\ell} \cdot b_{\ell k} \qquad \begin{aligned} i &= 1. \ . \ .NSAMP \\ k &= 1. \ . \ .NY \end{aligned} \qquad (6)$$

Figure 1 is the summary of the two block PLS algorithm using the
equation numbers.

Finally, the regression coefficient matrix S is calculated as
a function of A, B, C and $\underline{d}$.

$$S = \sum_{\ell=1}^{NCOMP} \underline{d} \cdot Z \cdot A \cdot B^T \qquad (7)$$

where Z initialized as an (NX * NX) identity matrix and in each term
is updated as

$$Z = Z - \underline{d}Z \ A \cdot C^T \qquad (8)$$

Properties of the PLS method

PLS gives a parallel solution of regression models for several response variables. Handling the response variables together, especially
in case of highly correlated y's , stabilizes the solution for the
regression coefficients, i.e. reduces the variance of the estimates.
PLS mitigates the colinearity problem (high correlation among
the predictor variables) by regressing the response variables on
orthogonal latent variables. In this respect the PLS regression is
similar to the principal component regression. Increasing the
number of components (number of latent variables) to the number of
predictor variables, the PLS solution converges to the least squares
solution. However, by removing some of the later latent variables,
which (similar to the principal components) describe only variance
due to noise, the variance of the regression coefficient estimates
can be reduced. The fit of the model becomes worse than that of the
least squares solution, but the predictive power of the model is
enhanced as variance due to random noise is omitted from the model.

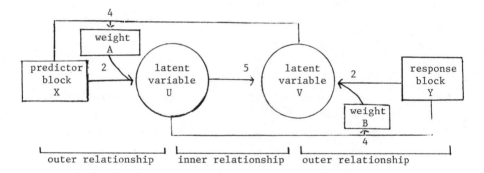

Figure 1. Two block PLS (numbers correspond to the equation numbers).

PLS (similar to ridge regression) trades bias for variance in case
of calculating fewer components (latent variables) than the number
of predictor variables.

The optimal number of components from the prediction point of
view can be determined by cross-validation (10). This method compares the predictive power of several models and chooses the optimal one. In our case, the models differ in the number of components.
The predictive power is calculated by a leave-one-out technique, so
that each sample gets predicted once from a model in the calculation
of which it did not participate. This technique can also be used to
determine the number of underlying factors in the predictor matrix,
although if the factors are highly correlated, their number will be
underestimated. In contrast to the least squares solution, PLS can
estimate the regression coefficients also for underdetermined
systems. In this case, it introduces some bias in trade for the
(infinite) variance of the least squares solution.

Comparison of PLS with the CMB and TTFA methods

The two mostly used statistical methods for calculating receptor
models are: CMB and TTFA.

The assumption of the CMB method is, that the mass concentration
of chemical element i, C_i, is a linear combination of the mass
fractions of the element i from source j, a_{ij}

$$C_i = \sum_{j=1}^{J} a_{ij} S_j \qquad i = 1. . .I \qquad (9)$$

The regression coefficients S_j are the source contributions, I is the
number of chemical elements, J is the number of sources. Note that
there is only one air sample and I>J has to be true to be able to
solve the regression by ordinary least squares. There are two problems. First, the predictor variables can be highly correlated.
Therefore, the solution for the regression coefficients, i.e. for the
source contribution S_j, is unstable (has high variance). Second, the
predictor variables are not error free. Therefore, their errors also
have to be included in the model. Solutions often used for the first
problem are ridge regression, which introduces bias to decrease the
variance of the estimated regression coefficients, or principal
component regression, that performs regression on the orthogonal
linear combinations of the predictor variables rather thant on the
variables themselves. The second problem can be solved by the
effective variance least-squares method, where the samples (chemical
elements) are weighted by

$$W_i = (\delta_{C_i}^2 + \sum_{j=1}^{J} \delta_{a_{ij}}^2 \cdot S_j^2)^{-1} \qquad (10)$$

Since the weights depend on the source contributions S_j, to be calculated, an iterative procedure is necessary.

The PLS solution for the CMB model has several advantages.
Instead of solving for one air sample at a time or for their average,

a solution for all the samples as a function of time or site can be achieved in one step. Therefore, the model can be extended as

$$c_{ik} = \sum_{j=1}^{J} a_{ij} \cdot s_{jk} \qquad \begin{array}{l} i = 1. \; . \; .I \\ k = 1. \; . \; .K \end{array} \qquad (11)$$

where the columns of S reflect the variance in time or site. K is the number of air samples. Including all the air samples in the same model enhances the stability of the estimates. PLS, similar to the ridge regression, trades bias for variance. However, determination of the optimal number of components by cross-validation is more straightforward than the choice of the ridge parameter k. The optimal number of PLS components chosen by cross-validation is an estimate for the number of the active source profiles. The effective variance weight scheme in Equation 10 can be extended to several air samples as

$$W_i = (\sum_{k=1}^{K} (\delta^2_{c_{ik}} + \sum_{j=1}^{J} \delta^2_{a_{ij}} \cdot s^2_{jk}))^{-1} \qquad (12)$$

and included to the PLS solution.

Determination of which potential polluting source is active is possible by including the sources stepwise and comparing the predictive power of the models with different source contributions by cross-validation.

The goal of the TTFA method is to estimate the number of sources, to identify them and to calculate their contribution from the ambient sample matrix C (chemical component concentrations i measured during sampling periods or at sampling sites k) using as little a priori information as possible. As a first step, an eigenvector analysis of matrix C is performed.

$$C = A' \cdot S' \qquad (13)$$

where A' is the eigenvector or loading matrix and S' is the score matrix. The literature distinguishes Q (correlation between samples) and R (correlation between chemical elements) mode factor analysis with or without centering (11).

In R mode the source profile matrix A is obtained and the source contribution matrix S is calculated from C and A. The Q mode analysis gives an opposite solution.

The number of significant eigenvectors is the estimate for the number of the active sources. However, the eigenvectors are not necessarily representative of the source profiles or source contributions. They must be linearly combined to form the source vectors. This is done in the second step by target transformation.

$$C = A' \cdot R \cdot R^{-1} \cdot S' \qquad (14)$$

where R is the rotation matrix. A' $\cdot$ R is the representation of the source profiles A and $R^{-1}S'$ of the source contributions S. R is determined by a series of least-squares fits of one test profile vector at a time on the eigenvector matrix A'.

A PLS solution similar to the TTFA approach can be obtained in one step when the air sample matrix X is used as predictor matrix and the source profile matrix A is used as response matrix.

$$a_{ij} = \sum_{k=1}^{K} c_{ik} \cdot s_{kj} \qquad \begin{array}{l} i = 1. \ . \ .I \\[1ex] j = 1. \ . \ .J \end{array} \qquad (15)$$

The number of active sources is estimated by cross-validation, i.e. it is the optimal number of PLS components. The latent variables of the PLS model would correspond to the eigenvectors of the TTFA model. The linear combination of the latent variables in the inner relationship gives the estimate for the source profiles. PLS calculates the orthogonal linear combinations and the rotation in one step. Also, it solves for all the sources in the same model. The effective variance weighting scheme (Equation.12) can be used also in this model to down-weight the chemical elements with high uncertainties.

The Quail Roost Data

A couple of years ago a workshop was organized to compare the performance of the various statistical methods applied for receptor model (12). To create an objective basis for the comparison of the different analyses, a synthetic data set was generated according to the following equation:

$$\tilde{c}_{ik} = \sum_{j=1}^{J} (\tilde{a}_{ij} - e_{ij}) \cdot s_{jk} + e_{ik} \qquad \begin{array}{l} i = 1. \ . \ . \ 20 \\[1ex] j = 1. \ . \ . \ 13 \\[1ex] k = 1. \ . \ . \ 40 \end{array} \qquad (16)$$

Both the air sample matrix C and the matrix of the potential source profiles A were perturbed by measurement error. In Set I only 9 sources were active, among which there was an unreported source. Set II was generated using all 13 profiles. These data sets are used to illustrate the performance of the PLS solution.

In Set I, cross-validation found 6 underlying components, instead of the true 9, because of the high correlation among certain profiles and one source contribution being below the noise level. In Set II the estimate of the number of the active sources is 11 vs. the true 13. Table I contains the estimated source contributions by different PLS models. The first row contains the true values of the regression coefficients. The values in the second through fifth row are the estimates from PLS models: 1) including the average of the 40 air samples in the response block, 2) the weighted version of 1), 3) including all 40 samples in the response block, and 4) the weighted version of 3). The negative contributions at the first source are due to the fact that the method is not

constrained so that it would result only in positive coefficients.
The estimate of the first source is highly biased because it has
very small contribution.

In Table II the sums of squared residuals (RSS) of Set I are
found calculated by the TTFA type model solved by PLS. All 13
potential profiles were predicted from the 40 air samples, while
in reality there were only 9 active. The first row contains the
RSS's from PLS models predicting one source profile at a time, the
second row from the PLS model predicting all the source profiles
simultaneously. From the difference of the RSS's between the
first nine and the last four profiles it is clear that in this
data set there were only nine sources active. These results are
intended only to illustrate what kind of information is provided
by the PLS solution.

Table I. Estimated source contributions by different PLS models

model \ sources	1	2	3	4	5	6	7	8	9
true	0.05	1.3	2.0	4.2	4.7	8.0	3.3	2.4	7.1
average	-0.30	1.3	0.77	4.7	5.3	7.2	4.4	3.3	7.0
average, weights	0.07	1.4	2.5	4.1	5.3	7.7	3.4	1.9	7.2
all 40	-3.6	0.9	3.0	5.1	6.2	7.4	4.3	1.9	6.2
all 40, weights	-1.1	0.8	2.3	5.0	6.0	7.5	4.2	2.3	7.0

Table II. Sum of squared residuals of the source profiles
predicted from Set I

model \ sources	1	2	3	4	5	6	7	8	9	10	11	12	13
one profile at a time	691	150	37	52	99	31	135	91	44	1504	11931	981	1003
all 13 profiles	918	181	53	50	90	103	210	40	53	2513	17805	1511	2104

Conclusion

In this paper the PLS method was introduced as a new tool in calculating statistical receptor models. It was compared with the two most popular methods currently applied to aerosol data: Chemical Mass Balance Model and Target Transformation Factor Analysis. The characteristics of the PLS solution were discussed and its advantages over the other methods were pointed out. PLS is especially useful, when both the predictor and response variables are measured with noise and there is high correlation in both blocks. It has been proved in several other chemical applications, that its performance is equal to or better than multiple, stepwise, principal component and ridge regression. Our goal was to create a basis for its environmental chemical application.

Literature Cited

1. Sjöström, M.; Wold, S.; Lindberg, W.; Persson, J.; Martens, H. Anal. Chim. Acta 1983, 150, 61.
2. Lindberg, W.; Persson, J.; Wold, S. Anal. Chem. 1983, 55, 643.
3. Bisani, M.L.; Faraone, D.; Clementi, S.; Esbensen, K.H.; Wold, S. Anal. Chim. Acta 1983 150, 129.
4. Frank, I.E.; Kalivas, J.H.; Kowalski, B.R. Anal. Chem. (1983), 55, 1800.
5. Frank, I.E.; Feikema, J.; Constantine, N.; Kowalski, B.R. J. Chem. Info. Comp. Sci. 1984, 24, 20.
6. Frank, I.E.; Kowalski, B.R. Anal. Chim. Acta 1984, 162, 241.
7. Frank, I.E.; Kowalski, B.R. Anal. Chim. Acta, in press.
8. Kowalszyk, G.S.; Gordon, G.E.; Rheingrover, S.W. Envir. Sci. Technol. 1982, 16, 79.
9. Hopke, P.K. The Application of Factor Analysis to Urban Aerosol Source Resolution, ACS SYMPOSIUM SERIES, No. 167, 1981.
10. Stone, M. J. Roy. Stat. Soc. 1974, Ser. B, 36, 111.
11. Malinowski, E.R.; Howery, D.G. "Factor Analysis in Chemistry", J. Wiley and Sons, New York, 1980.
12. Stevens, R.; Dzubay, C.,; Lewis, C.; Currie, L.; Johnson, D.; Henry, R.; Gordon, G.; Davis, B. "Mathematical and Empirical Receptor Models: Quail Roost II", U.S. E.P.A. report, 1982.

RECEIVED July 17, 1985

Author Index

Subject Index

A

280

Production by Hilary Kanter
Indexing by Deborah H. Steiner
Jacket design by Pamela Lewis

Elements typeset by Hot Type Ltd., Washington, D.C.
Printed and bound by Maple Press Co., York, Pa.